T0227582

THE
QUARK
MACHINES

Sir John Adams (1920–84) personified Europe's post-war achievements in physics. With international collaboration under his inspired leadership, an old continent impoverished by war was re-established in the front rank of world science.

THE QUARK MACHINES

HOW EUROPE FOUGHT THE PARTICLE PHYSICS WAR

GORDON FRASER

CRC Press
Taylor & Francis Group
Boca Raton   London   New York

CRC Press is an imprint of the
Taylor & Francis Group, an informa business

CRC Press
Taylor & Francis Group
6000 Broken Sound Parkway NW, Suite 300
Boca Raton, FL 33487-2742

First issued in hardback 2017

© 1997 by Taylor & Francis Group, LLC
CRC Press is an imprint of Taylor & Francis Group, an Informa business

No claim to original U.S. Government works

ISBN 13: 978-1-138-40635-3 (hbk)
ISBN 13: 978-0-7503-0447-4 (pbk)

This book contains information obtained from authentic and highly regarded sources. Reasonable efforts have been made to publish reliable data and information, but the author and publisher cannot assume responsibility for the validity of all materials or the consequences of their use. The authors and publishers have attempted to trace the copyright holders of all material reproduced in this publication and apologize to copyright holders if permission to publish in this form has not been obtained. If any copyright material has not been acknowledged please write and let us know so we may rectify in any future reprint.

Except as permitted under U.S. Copyright Law, no part of this book may be reprinted, reproduced, transmitted, or utilized in any form by any electronic, mechanical, or other means, now known or hereafter invented, including photocopying, microfilming, and recording, or in any information storage or retrieval system, without written permission from the publishers.

For permission to photocopy or use material electronically from this work, please access www. copyright.com (http://www.copyright.com/) or contact the Copyright Clearance Center, Inc. (CCC), 222 Rosewood Drive, Danvers, MA 01923, 978-750-8400. CCC is a not-for-profit organization that provides licenses and registration for a variety of users. For organizations that have been granted a photocopy license by the CCC, a separate system of payment has been arranged.

Trademark Notice: Product or corporate names may be trademarks or registered trademarks, and are used only for identification and explanation without intent to infringe.

Visit the Taylor & Francis Web site at
http://www.taylorandfrancis.com

and the CRC Press Web site at
http://www.crcpress.com

CONTENTS

ACKNOWLEDGMENTS AND DISCLAIMERS

In some respects, the sociology of elementary particle physics resembles a game of soccer, with rival but nevertheless friendly teams striving to score research goals. Once scored, the research goals become monuments, while their scorers achieve fame and are showered with honours. Strikers command higher transfer fees than defenders, but goals do not come about without assiduous preparatory teamwork, unremitting trial and error, and skilful play. Just as foreplay inevitably gets overlooked and is edited out of television action replays, so it is in this book. Many scientists try to be meticulous in crediting research, conscientiously documenting all the intricate passes which lead to a particular goal, and without which the goal would most likely never have come about. Tracing the origins of a goal is a laudable aim but does not necessarily make compelling reading.

Another citation dilemma arises from one of the basic themes of the book—the dependence of elementary particle physics research on larger and larger installations, demanding scientific teamwork on an unprecedented scale. Over decades, these teams have grown to be measured in tens, then hundreds and now thousands of specialists. While such teamwork means that many people are thrust into key roles, they unfortunately cannot all be mentioned. This book acknowledges the teams, but names have been limited mainly to those whose fate was to be honoured anyway, or to those who lived at a time when collaborations were more intimate and science was less impersonal.

Trying to paraphrase what can only be expressed with mathematical precision is another hazard, an intellectual assault course which runs the gauntlet of heavy dogmatic firepower. Favouring ease of assimilation by the uninitiated demands some forbearance on the part of the initiated, with whom I shall not be popular.

The book also has a distinct transatlantic flavour and sidesteps much of what happened in parallel in the then Soviet Union and in Japan.

Thanks are due to the many people, again too numerous to mention, at CERN and elsewhere, who have patiently tried to impart their wisdom. Jim Allaby, David Cline, Don Cundy, Jasper Kirkby, Maurice Jacob, Kjell Johnson, Phillipe Lebrun, Keith Potter, Alvaro de Rùjula, Johnathan Rosner, Roy Schwitters, Konrad Spiegel, Jack Steinberger and Ted Wilson kindly read draft extracts and suggested improvements but are in no way responsible for any remaining deficiencies. I am also grateful to Valentine Telegdi for his erudition and historical awareness, and to Egil Lillestøl and Frank Close, who advised me when my head was under water. Thanks are also due to Peter Binfield, Katie Pennicott, Robin Rees and Jim Revill at Institute of Physics Publishing for transforming a vague initial idea into this book.

Many thanks are also due to CERN photographer Patrice Loïez for his continual and enthusiastic cooperation. Ben Fraser and Markus Audria supplied the line diagrams.

Gordon Fraser
Divonne-les-Bains
April 1997

About the author
With a first-class degree in physics and a doctorate in the theory of elementary particles from London's Imperial College, Gordon Fraser began his career in research but soon moved into technical journalism. Returning to laboratory work, this time as Information and Press Officer at the Rutherford Laboratory in the UK, he transferred to scientific publications at CERN, the European Laboratory for Particle Physics in Geneva, where for the past eighteen years he has edited *CERN Courier*, the international monthly news magazine of high-energy physics, whose annals have been a prolific source of material for this book. With Egil Lillestøl and Inge Sellevåg, he is the author of *Search for Infinity* (London: Mitchell Beazley, 1994; New York: Facts on File, 1995), which has been translated into nine languages.

Photographic credits are as follows: frontispiece, figures 2–9, 12, 14, 15, 17–19, 21, 23, 24, 27, 28, 33, 36, 39–44, 46, 51 and 53, CERN; figure 1, NASA; figure 10, Harvard University, Courtesy AIP Emilio Segré Visual Archives; figure 11, UKAEA plc; figures 16 and 20, Brookhaven National Laboratory; figure 22, Lawrence Berkeley Laboratory; figure 30, G Groote, Bad Honnef; figure 34, G Fraser; figures 35 and 45, Fermilab Visual Media Service; figures 47–50, Supercollider Photographic Services; figure 52, M Goidadin.

GLOSSARY

This ready reference list of some key words and personalities is not meant to be exhaustive. Words in *italics* refer to other entries in the glossary. For full details, see the index.

Adams, Sir John British engineer (1920–84). Designer of *CERN*s major *synchrotrons*, and several times Director General of *CERN*

AGS (Alternating Gradient Synchrotron) Brookhaven particle accelerator; commissioned 1960

Amaldi, Edoardo Italian physicist (1908–89); one of the founders of *CERN*

Antimatter A variant of matter in which all additive quantities (such as electric charge) have opposite values to those of ordinary matter; can be understood as ordinary matter moving backwards in time

Antiparticle A particle of *antimatter*

Asymptotic freedom Quantum *field theories* in which interactions become stronger as the distance between particles increases

Auger, Pierre French physicist (1899–1994). One of the founders of *CERN*

Bohr, Niels Danish physicist (1885–1962); long-time figurehead of European science

Brookhaven US National Laboratory on Long Island, New York; founded 1947

CERN (originally from Conseil Européen pour la Recherche Nucléaire) European Laboratory for Particle Physics based in Geneva, Switzerland; founded 1954

CESR (Cornell Electron Storage Ring) Electron–positron *collider* at Cornell. Commissioned 1979

Charm Charge-like quantity carried by certain *quarks*, analogous to *strangeness*

Collider Ring or rings in which contra-rotating beams of high energy particles collide with each other

Colour The charge-like quantity carried by the *gluons* of the inter-*quark* force

Cyclotron Particle accelerator invented by Ernest *Lawrence* in 1930 in which electrically charged particles rotating in a magnetic field spiral outwards as they receive periodic boosts of electrical energy

1

DESY (Deutsches Elektron Synchrotron) German National Laboratory in Hamburg. First *synchrotron* operational 1964

DORIS (DOppel RIng Speicher) Electron–positron *collider* at *DESY*. Commissioned 1974

Dubna Russian laboratory north of Moscow; founded 1956

ECFA (European Committee for Future Accelerators) A user forum of high-energy physicists

Electron Atomic particles surrounding the nucleus; carry a negative electric charge; no inner *quark* structure

Electronvolt (eV) The energy gained when an electron passes across a potential difference of 1 V

Electroweak interaction The combined effects of electromagnetism and the *weak interaction*

Fermi, Enrico Italian physicist (1901–54); emigrated to the USA after winning the 1938 Nobel prize

Fermilab Fermi National Accelerator Laboratory near Chicago. Established 1966; site of several large *synchrotrons*

Feynman, Richard US physicist (1918–88); winner of Nobel prize 1965; pioneer of *quantum electrodynamics*

Field theory A mathematical accounting system for keeping track of particle interactions

Fission The ability of some large unstable nuclei to split into sizable chunks, rather than spitting out small *radioactive* particles

Gell-Mann, Murray US physicist; winner of Nobel prize 1969; inventor of the *quark* picture.

GeV Giga*electronvolt*; 10^9 eV

Glashow, Sheldon US physicist; winner of Nobel Prize 1979; pioneer of the *electroweak* picture.

Gluon Particle carrying the inter-*quark* 'colour' force

HEPAP US High Energy Physics Advisory Panel

HERA (Hadron-Elektron-Ring-Anlage) Electron–proton *collider* at *DESY*; commissioned 1991

Higgs mechanism Delicate directional effect in the vacuum which plays a major role in the *electroweak* process; origin unknown

ICFA   (International   Committee   for   Future   Accelerators)   Makes recommendations on intercontinental physics collaboration

ISABELLE   Proton–proton *collider* planned at *Brookhaven*. Subsequently cancelled and later resurrected as the Relativistic Heavy-Ion *Collider*; scheduled operation 1999

ISR   (Intersecting   Storage   Rings)   World's first proton–proton *collider* commissioned at *CERN* in 1971; closed 1984

IUPAP   International Union of Pure and Applied Physics

Kaon   (or   K   meson)   Unstable particle created in subnuclear collisions; composed of a *quark* and an antiquark and carrying *strangeness*

Lawrence, Ernest O   US physicist (1901–58); inventor of the *cyclotron*

LEP (Large Electron Positron Ring)   *CERN*'s 27 km electron–positron *collider*; commissioned 1989

LHC (Large Hadron Collider)   Proton–proton *collider* under construction in *CERN*'s 27 km *LEP* tunnel; planned completion 2005

Manhattan   Project   US Second World War atomic bomb development programme

MeV   Mega*electronvolt*; 10^6 eV

Muon   A heavy variant of the *electron*; involved in *weak interactions*

Neutral current   Variant of the *weak interaction;* involving no permutation of electric charge; carried by the *Z particle*

Neutrino   Almost invisible particle involved in *weak interactions*; uncharged

Neutron   Electrically neutral subnuclear particle; made of three *quarks*

Oppenheimer, J Robert   US physicist (1904–67); scientific director of the *Manhattan Project*

Partons   Point-like particles deep inside protons and neutrons; now known to be quarks

PEP   Electron–positron *collider* at *SLAC*. Commissioned 1980

PETRA (Positron Electron Tandem Ring Accelerator)   Electron–positron *collider* at *DESY*; commissioned 1978

Pion (or pi meson)   Unstable particle created in subnuclear collisions; composed of a *quark* and an antiquark; discovered in 1947

Proton   Subnuclear particle carrying positive electric charge and made of three *quarks*

Quantum chromodynamics   The quantum *field theory* of the '*colour*' force between *quarks*

Quantum electrodynamics   The quantum *field theory* of electricity and magnetism

Quark   Ultimate constituent of nuclear matter; carries multiples of one third of an electric charge

Rabi, Isidor   US physicist (1898–1988); winner of Nobel prize 1944; played a key role in setting up major post-1945 physics laboratories

Radioactivity   Spontaneous disintegration of unstable nuclei

Renormalization   The tractability of a quantum *field theory*; some theories produce infinities which make meaningful calculations impossible

Rubbia, Carlo   Italian physicist; winner of Nobel prize 1984 for the discovery of *W* and *Z particles*; Director General of *CERN* 1989–93

Salam, Abdus   Pakistani physicist (1926–96); winner of Nobel prize 1979; pioneer of the *electroweak* picture

Serpukhov   Russian laboratory south of Moscow; site of 70 GeV proton *synchrotron*

Schwinger, Julian   US physicist (1918–94); winner of Nobel prize 1965; pioneer of *quantum electrodynamics*

SLAC (Stanford Linear Accelerator Center, California)   Site of a 2 mile linear electron accelerator; commissioned 1966

SPEAR (Stanford Positron–Electron Asymmetric Rings)   Electron–positron *collider* built at SLAC in 1973

SPS (Super Proton Synchrotron)   *CERN*'s 7 km *synchrotron*; commissioned in 1976

SSC (Superconducting Supercollider)   Construction of this 87 km US proton *collider* began in Texas in 1992; the project was cancelled the following year

Standard Model   The current picture of *quarks* and *leptons* interacting through the *electroweak* and *colour* forces

Strangeness   Quantity resembling electric charge carried by the component quarks in certain unstable subnuclear particles

Synchrocyclotron   High-energy variant of the *cyclotron* in which the electric or magnetic fields are modulated to take account of the effects of relativity

Synchrotron   Particle accelerator in which bunches of charged particles rotate in an annular tube under synchronized electric and magnetic fields

TeV   Tera*electronvolt*; 10^{12} eV

Tevatron   *Fermilab*'s highest-energy *synchrotron*

Top   The heaviest *quark*

W particle   Electrically charged carrier of the *weak interaction*

Weak interaction   The agent which transforms one kind of *quark* to another

Weinberg, Steven   US physicist; winner of Nobel prize 1979; pioneer of the *electroweak* picture

Weisskopf, Victor   Austrian physicist; Director General of *CERN* 1961–5

Wideroë, Rolf   Norwegian physicist and engineer (1902–96); inventor of the linear accelerator

Wilson, Robert Rathbun   US engineer and physicist; founding Director of *Fermilab*

Z particle   Electrically neutral carrier of the *weak interaction*

SUBNUCLEAR LIFTOFF

'Those who have chosen to worship men as gods ... are greatly mistaken, since even if a man were as large as our Earth he would seem like one of the least of the stars which appear as but a speck on the Universe.'

Leonardo da Vinci

The rapid scientific advances of the twentieth century have led to a new awareness of Man's position in the Universe. Not that long ago, scientific dogma reflected an arrogant belief that human existence occupied some key role in the order of things. As investigations progressed and understanding advanced, this cultural arrogance melted into an increasing awareness of a vast cosmic sequence. From our fragile perch, we gaze up towards the enormity of outer space and the cosmos and peer down to an atomic and subatomic world which is equally distant from our own experience.

The conquest of outer space is one of the twentieth century's major achievements, symbolized by carefully orchestrated countdowns when huge rocket motors are ignited, the earth shakes to the thunder of thousands of tonnes of thrust, and titanic space vehicles teeter hesitantly off the ground before eluding the clutch of gravity and soaring majestically aloft. Such liftoffs are not themselves discoveries. They are points of departure, marking the beginnings of new phases in the continual quest to push back the frontiers of knowledge. Only after the successful launch of the rocket can the payload, equipped with its sensitive instruments, begin its mission of exploration.

The conquest of the 'inner space' of the subatomic microcosmos is no less impressive, but from our perch higher up on the cosmic sequence is much less visible. To probe inner space, satellites are replaced by high-energy beams of subatomic particles, such as protons, each so small that lining up ten billion (where 1 billion$=10^9$) of them (as many grains of sand as it would take to fill a large box) would only stretch just one hair's breadth. For this research the mighty rockets which launch satellites for space exploration are parallelled by huge magnetic racetracks in which particle beams are hurled round and round. These tools to explore the subatomic world (hence 'atom smashers') are in principle no less impressive than their rocket counterparts but make less noise and do not themselves move. Subatomic countdowns, marking the launch of a new machine to open up the exploration of inner space, attract less media attention.

One historic such countdown took place in January 1971 at CERN (see pages 116–118), near Geneva, Switzerland, the European subnuclear equivalent of

6

Figure 1: *Liftoff for outer space. Smoke and thunder on 24 April 1990 when, with the Hubble Space Telescope aboard, the Space Shuttle* Discovery *leaves the launchpad at the Kennedy Space Center.*

NASA's Cape Canaveral (now Cape Kennedy). CERN had just completed construction of a new machine, the Intersecting Storage Rings (ISR). This novel device had been designed to store beams of subnuclear protons in two interlaced rings 300 m in diameter, so that the circulating protons would clash together, with the particles repeatedly brought into collision at the eight points where the rings intersected. Elsewhere, physicists had succeeded in colliding beams of electrons, but protons were more difficult to master. Where subnuclear scientists had previously relied on research probes which hurled a single high energy proton beam into an inert stationary target, the ISR was an ambitious new departure in research. 'Too ambitious,' said some. Such circulating beams could be easily destroyed by chance encounters with residual gas molecules. The ISR rings therefore had to contain an almost perfect vacuum—two tubes of outer space. The technology to ensure such a high vacuum had to be developed from scratch, but there was another unknown. After the rings had been painstakingly pumped free of air, they would then be filled with protons. When they clashed together, the ISR beams might jostle each other too much for comfort and refuse to circulate smoothly.

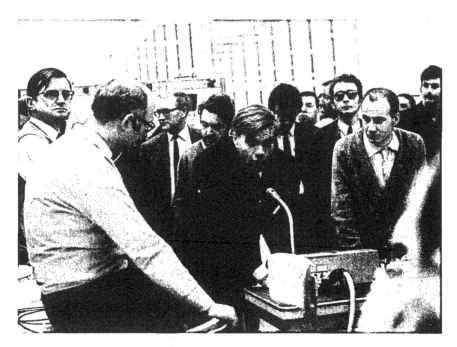

Figure 2: *Liftoff for inner space. No fireworks on 27 January 1971 when Kjell Johnsen announced that the first subnuclear particle collisions had been achieved in the new ISR at CERN.*

In December 1970, the ISR's Ring 1 had been fed with protons for the first time, and had attained a world record proton current of 1.5 A (meaning that the ring held 30 million million circulating protons). Electric currents are conventionally made of electrons (subatomic particles), flowing along a wire. The ISR was instead providing a sizable current of protons (subnuclear particles). The goal for 1971 was to store protons in both rings at the same time so that the intersecting proton beams would collide. The control room held its breath as the proton current in Ring 1 was steadily increased to 930 mA, almost 1 A of protons, not as many as in the initial December test, but enough for the next phase. After a few minutes there was an audible groan when the stored proton current inexplicably dropped to 587 mA. How had so many protons been lost? If this still useful current would continue to fade at that rate, the protons could not be considered as being stored and there would be no chance of success. However, two hours later, the Ring 1 current was still 587 mA. After the initial hiccup, the decay rate per second of the current was only five parts in a hundred million, implying that a useful proton current could be maintained for several months without needing to be replenished. Ring 1 of the ISR had been launched and was in orbit.

Building up the Ring 1 current had required many injection shots of protons. Meanwhile a single shot of protons had been fired into ring 2, giving a modest 15 mA. The two currents, circulating in opposite directions, were crossing each other eight times on each ISR revolution. The physicists had feared that the intersecting beams would get in each other's way, destroying the carefully established beam conditions and preventing the ISR from attaining its objective of coasting beams. But no! The tiny proton current in Ring 2 held fast. Liftoff! The protons were both colliding and continuing to circulate. A control room announcement reporting that the world's first collisions between proton beams were taking place was greeted by a polite ripple of applause. Multiplied by the number of protons, even just in Ring 2, the polite applause became thunder.

It was not a scientific discovery. The ISR was to win no Nobel prizes. Instead, it was a point of departure. A new phase in the exploration of inner space had begun but, while the ISR liftoff was a major success, its initial instrumentation payload was inadequate to explore the new horizons that it opened up. The ISR liftoff was not the last time such subatomic thunder was to reverberate at CERN, but on the next and subsequent occasions researchers had learned their lesson and took care that the instrumentation payload (the subnuclear particle detectors) rose to the challenge.

Some forty years before this ISR liftoff, in January 1933, Adolf Hitler, as Chairman of the National Socialist German Workers' Party, became Chancellor of Germany. The following month, the burning of the Reichstag, Germany's parliament buildings, further fuelled Nazi malevolence. In April 1933, the civil service 'Restitution' act stipulated that 'civil servants not of Aryan ancestry are to be placed in retirement'. Antisemitism became legal. The civil servant category included university staff, and a wave of intellectual emigrants moved westwards, many taking advantage of the traditional immigrant hospitality of the USA.

Over half a century later, on 9 November 1989, the Berlin Wall fell and East Germany ceased its imposed isolation. What had begun in 1933 in Germany had run its course. However, much pent-up energy had yet to be dispelled. Within a few years, the Soviet Union was no more and the Cold War between East and West melted as quickly as it had frozen. In a politico-economic climate which had been inconceivable just a few years before, on 4 October 1993, US President Bill Clinton signed a bill to halt the construction of what would have been the world's largest scientific instrument—the Superconducting Supercollider (SSC), a giant physics machine being built for an 87 km underground tunnel south of Dallas, Texas.

This last enactment, disconnected from what happened in 1933 by the unpredictable twists and turns of more than half a century of momentous history, marked a turning point in US science, the end of an epoch during which the country had dominated subatomic physics, the quest to find and understand the smallest particles of matter. For some fifty years, this branch of physics had been as American as baseball and bubble gum. During this period of US pre-eminence, the ISR liftoff had marked Europe's hesitant re-emergence on the subatomic physics stage, although not yet as a major player.

The quest to understand matter on the smallest scale had emerged from the philosophy of the Ancient Greeks. With the unimaginably small-scale structure of matter out of reach, this branch of science began as speculative philosophy. The Renaissance saw the development of the first instruments, namely the microscope and the telescope, which enabled scientists to study the very small and the very large. Horizons immediately widened and physics increasingly became an experimental science. New ideas swept aside old dogma. To become accepted, it was not sufficient that a new theory had a staunch and celebrated advocate. Instead, the ultimate test became verification by experiment—trial by ordeal. With increased technological prowess, science continued to make great progress, but the innovative instrumentation of the twentieth century ushered in a golden age. Fresh discoveries begged for explanation. New theoretical pictures demanded experimental verification. The discovery of atomic structure and of the atomic nucleus on the experimental side went hand in hand with the development of the new conceptual pictures of relativity and quantum mechanics. In the first thirty years of the twentieth century, physicists realized that Nature operated on the subatomic scale in bizarre ways that were far removed from everyday experience.

Virtually all this rapid development took place in Europe, in ancient universities whose illustrious traditions had instilled intellectual curiosity, tempered with rigour and discipline. As a young nation, the USA eagerly sought to absorb this new culture, and a period of work at a major European university was mandatory for aspiring US physicists. However, the unaccustomed rigour and discipline of European science was not an easy road for American researchers to follow. As with teenage rebels, unfettered imagination and ambition are difficult to reconcile with caution and discipline.

Hitler's 1933 decisions and the consequent influx of top-flight European intellectual immigrants accomplished within a few years what might otherwise have taken decades. With Albert Einstein as figurehead, a Who's Who of subnuclear physicists crossed the Atlantic. As these scientists grew in stature after establishing themselves in initially unfamiliar new positions, the spark of a fresh nuclear discovery lit a mighty blaze. The elucidation of nuclear fission on the eve of the Second World War was a springboard for nuclear physics. The skills and knowledge of the new immigrants, coupled with the economic might of the USA, together created one of the greatest single undertakings the planet has ever known—the Manhattan Project, a race against the clock to develop an atomic bomb. The race beat the clock, and physics had been transformed from an intellectual backwater into a strategic weapon. The physicists who had developed the bomb and hastened the war to its conclusion became national figures.

Their duty done and with a heavy collective conscience, many physicists deserted their weapons laboratories and rushed back to their peacetime research, but no sooner had the clouds of the Second World War dispersed than a new threat loomed. With the delicate balance of this strange new Cold War, fresh national interests had to be safeguarded. The Manhattan Project's nuclear physicists were no longer themselves weapons builders but had become instead the teachers who would have to train future generations of weapons builders. To do physics was patriotic.

Nazi Germany had destroyed a nation's scientific heritage almost overnight, but the USA built its own almost as quickly. The wartime physicist heroes were granted their every wish and their university research benefited. The result was a scientific bull market which brought a wealth of new discoveries and fresh understanding, results and insights arriving head over heels, almost too fast to be absorbed. To study inner space and understand how the atomic nucleus works, the physicists needed huge machines, particle accelerators which take beams of subatomic particles to very high energies. With these twentieth-century 'microscopes', physicists could illuminate the interior of the atom and in turn even the nucleus itself and peer inside. As with the exploration of outer space, building and exploiting these huge scientific machines required unprecedented enterprise and effort, on a level to match that which had been seen for the Manhattan Project. Soon the USA had an effective world monopoly on particle-accelerating power.

Probed in depth with these new machines, the atomic nucleus was found to be more complicated even than the atom which surrounded it. On a new scale of cosmic intricacy, as small again compared with the nucleus as the nucleus compared with the atom, the protons and neutrons which make up atomic nuclei were found to contain a new layer of fundamental matter, the 'quarks'. Earlier in the century, the staid European intellectual approach had introduced elegant subnuclear words such as 'electron' and 'proton'. Americans less impressed by classical culture instead coined nonsense words such as 'strangeness' and 'quark' to describe new concepts. Many scientists were sceptical. It was as though a new solar system planet had been found and christened 'Mickey Mouse', but the unaccustomed flippancy of the new subnuclear nomenclature was not to be taken lightly. It cloaked a new depth of insight, a radical reappraisal of dogma which itself had hardly had time to gain acceptance.

With this postwar subnuclear research boom in the USA, it looked like Europe might have missed the boat. The continent which had nurtured basic physics looked to have lost its birthright, but new European feelings were emerging. In the aftermath of the Second World War, Europe, the culprit of two wars that had gone on to rock the entire planet, hastened to present a new unified front. On a world scale, the new United Nations (UN) sought to pre-empt war by stimulating collaboration, establishing new specialized agencies. International- ism was the cry, in politics, industry, trade, culture and science. In the forums created to discuss such collaborations, new voices were heard, and new pressures felt. One pilot scheme in European collaboration launched in the early 1950s was different. Instead of being a platform for discussion and bargaining, it was goal oriented. This was CERN, an international physics laboratory. Almost from the outset, it set itself objectives, clearly delineated goals in subnuclear research.

In 1953, CERN's initial goal was so startlingly ambitious that it looked almost foolhardy, namely the Europeans set out to challenge the Americans at their new game and to build the world's largest particle accelerator. However, the European machine duly provided its first beams in November 1959. The bold experiment in international collaboration had worked, demonstrating the effectiveness of scientific teamwork on the large scale. When specific projects were addressed, international collaboration and teamwork proved very effective. After this initial success, Europe witnessed a series of successful subnuclear liftoffs. With the long time-scale needed to bring these projects to fruition, as soon as one project neared launchpad readiness, the focus switched to the next objective. Europe, with CERN as its figurehead, became a modest competitor in the USA-led race.

After being launched, the instruments deployed in subnuclear space have to seek and find. It looked for a time as though the Europeans could design and launch fine new tools to explore inner space, but their harvest of discovery was disappointing. It was the Americans who walked off with the research prizes. Slowly the Europeans learned how to be bold, to discard preconceptions and to go for long shots instead. At the same time, they retained their tradition of meticulous care. In the 1970s and 1980s, the research fruits began to arrive. Slowly the physics pendulum which had lurched westwards in 1933 began to edge back east, at first imperceptibly and then with gathering momentum.

In the late 1980s, with the arrival of glasnost and perestroika in the former Soviet Union, the fall of the Berlin Wall and the sudden evaporation of the Cold War, the motivation which underpinned the traditional US investment in nuclear physics suddenly evaporated. The result was a drastic reappraisal of objectives, acted out against a vivid contemporary backdrop of decreased public spending. In 1993, the 87 km Superconducting Supercollider (SSC), the world's largest machine, a ring of magnets large enough to encircle a major city, being prepared as the flagship of US subnuclear physics for the dawn of the twenty-first century, was unceremoniously scrapped. It was a staggering setback for the glorious tradition of post-war US subnuclear physics.

While welcoming the return of the physics pendulum, Europe, where the austerity of the 1990s was felt just as keenly, also reeled under the financial weight of its new responsibilities. In Europe, global pressures were reinforced by stringent limits in public spending imposed by the 1992 Maastricht treaty to prepare the way for European financial union. While the USA had chosen to sacrifice big projects on the financial altar, Europe preferred across-the-board cuts to limit national deficits. Projects such as CERN which had vividly displayed the effectiveness of European collaboration did not escape these financial restraints and had to make their own indirect contribution towards the next step in European unity. The financial winds buffeted the physics pendulum in its path. Once more science was the victim of contemporary issues.

EXODUS

For physics, the twentieth century jumped the gun when on 8 November 1895 in Würzburg, Germany, Wilhelm Röntgen had a sudden idea for an experiment with a cathode-ray tube. Covering the tube in black paper and darkening the room, he was surprised to see a fluorescent screen begin to glow. Some kind of radiation was penetrating the black paper. When Röntgen went public with his discovery in January 1896, the news of this 'x'-radiation triggered the science of subatomic physics.

In 1901, Röntgen won the first Nobel prize for physics for his 1895 x-ray discovery. In 1907, Albert Michelson of Chicago received the Nobel prize for his precision work in optical spectroscopy and metrology, the first American to receive the coveted prize. Eighteen years passed before the next US citizen, Robert Millikan of the California Institute of Technology in Pasadena, received the Nobel physics prize for his measurement of the charge of the electron and for his work on the photoelectric effect. This modest early US Nobel achievement is in stark contrast with more recent times when, over the twenty years from 1974 to 1994, 27 Americans won the Nobel physics prize compared with 16 Europeans and four from elsewhere in the world.

Another mirror which reflected physics fashion was the Solvay conference, first held in 1911. The Belgian chemist Ernest Solvay (1838–1922) had discovered a new process for manufacturing sodium carbonate on an industrial scale to feed the glass and soap industries. Solvay's wealth, like that of Alfred Nobel, was channelled into philanthropic projects, including the support of science, and the Solvay meetings provided a forum where the implications of revolutionary new physics ideas were debated. There was no American delegate at the 1911 Solvay meeting.

How did the USA rise so rapidly to physics fame in the latter half of the twentieth century? While it naturally took time for the young nation to achieve scientific maturity, there was one event which catalysed US pre-eminence in physics. The instigator of the event was Adolf Hitler.

During the early part of the century, experimental research in nuclear physics was dominated by the Cavendish Laboratory, Cambridge, first under J J Thomson, who won the Nobel physics prize in 1906, and then under Ernest Rutherford. Rutherford invented the science of nuclear physics and the Cavendish Laboratory flourished as its world focus. A tenacious experimentalist,

14

Figure 3: *Ernest, Lord Rutherford (1871-1937) invented the science of nuclear physics. A meticulous researcher, he did not let discoveries inadvertently slip from his grasp and was able to make remarkable achievements with unsophisticated apparatus.*

Rutherford had arrived at Cambridge from New Zealand as a student in 1895 with a scholarship for further study. The scholarship, awarded to a New Zealand student only every few years, had initially been earmarked for a young chemist, who decided at the last minute to get married and to remain in New Zealand. The scholarship passed to Rutherford, who, because of a change in university regulations, was able to use it at the Cavendish Laboratory under J J Thomson. Rutherford had only just managed to scramble aboard.

In New Zealand, Rutherford had carried out experiments on radio telegraphy and brought his prototype transmitter to Britain. In 1895, other scientists, including Guglielmo Marconi in Italy and Oliver Lodge in Britain, were also experimenting with radio telegraphy but, according to J J Thomson, Rutherford briefly held the world record for radio telegraphy transmission. However, on learning of Röntgen's dramatic x-ray discovery, Thomson mobilized his laboratory to attack this new research. Rutherford was ordered to drop radio telegraphy and to investigate atomic physics instead. Taking this abrupt change in research direction in his stride, Rutherford went on to win the Nobel prize in

1908 for his historic work on radioactivity, one year before Marconi shared the Nobel physics prize for his work on radio telegraphy. Rutherford's Nobel prize was in fact for chemistry, described by him as an 'instantaneous transmutation'. Paradoxically Rutherford, one of the leading figures of twentieth century physics, with a string of major discoveries to his credit, was never awarded a Nobel prize for physics.

After his introduction to subatomic physics at Cambridge, Rutherford moved to McGill University, Montreal, where he carried out his Nobel-prize-winning work on radioactivity, and then in 1907 moved back to the UK, to Manchester University. Here he made the key discovery of the nuclear atom, showing that most of the atom was empty space and its key properties concentrated in a tiny central core, and began the study of the nucleus itself, showing that it contained individual particles: protons. In 1919, Rutherford, already famous, succeeded J J Thomson as the Director of the Cavendish Laboratory, Cambridge. Over the next thirteen years, Rutherford's school at the Cavendish revealed the first nuclear transformations, discovered the neutron and developed the experimental techniques. Patrick Blackett became a virtuoso of C T R Wilson's cloud chamber, and John Cockcroft and Ernest Walton developed the first particle accelerator to achieve physics results.

These experimental techniques required only modest technology, even when judged against the standards of the time. Throughout his career, Rutherford maintained an approach to his own science that was unsophisticated but incisive. Apparatus remained modest and inexpensive, and the laboratory was traditionally locked at six in the evening. Only occasionally were experiments given a special dispensation to run longer hours. The evenings, stipulated Rutherford, were for reading and thinking.

A large man with a booming voice, Rutherford came from a colonial background and was either insensitive to or not intimidated by the subtleties of the British class system. His students and associates were selected on their ability alone. His approach was simple straightforward experimentation, leaving no stone unturned, with meticulous investigation of all possibilities before moving on. A classic example of his perseverance was his discovery of the atomic nucleus with Hans Geiger and Ernest Marsden at Manchester in 1909. At Rutherford's suggestion, Geiger was patiently investigating how alpha particles were deflected as they passed through a thin sheet of solid material. In 1908, Rutherford had discovered that alpha particles are helium atoms from which the electrons have been removed. With the human eye the only sensitive instrument available, monitoring alpha particles required a valiant effort. Using a small telescope to intercept the flashes produced when alpha particles hit a scintillating screen, and painstakingly covering the far side of the target sheet with the telescope patch by patch, Geiger and Rutherford mapped the scattering of alpha particles. Because such heroic observations were so demanding on the experimenters, it was perhaps natural that the laboratory was closed in the evening. Electronics was still in its infancy and apparatus could not be left to run itself, with supervisory researchers working shifts.

Despite the fact that most of the alpha particles went more or less right through the sheet, gently swerving past the atoms, Rutherford's special qualities showed when he insisted that all possibilities had to be covered, even the unlikely ones, and ruled that somebody should look in the backward hemisphere too. Marsden dutifully did so and, to everyone's astonishment, including Rutherford's, alpha-particle flashes were seen coming out backwards; a few alpha particles had hit something solid and bounced out along the way that they had come.

It took Rutherford the best part of a year to understand what had happened. Eventually he realized that most of the mass of the atoms in the target sheet was concentrated in tiny nuclei at the centre of the atoms, while the rest of the atom was empty space, populated by light electrons. Thus, while most of the alpha particles encountered very little, a few met a very solid obstacle. Suggesting that such an experiment should have been carried out in the first place, even when nothing special was expected, vividly displays Rutherford's thoroughness. Many scientific discoveries have been missed because the investigators, engrossed in an immediate goal, overlooked a discovery opportunity. In many major physics discoveries, someone had almost been there before. In the late nineteenth century, the British scientist William Crookes did everything but discover x-rays. Rutherford did not let discoveries inadvertently slip from his grasp and, when something unexpected did occur, had the tenacity and insight to follow it through.

Rutherford was the right man at the right time. Although he was a tenacious experimenter, his experiments were simple and straightforward, and it was remarkable that so many important discoveries could be made using such modest means. At a time when this aspect of science was developing so quickly, Rutherford had to remain receptive to new ideas. Although he has been portrayed as headstrong and resistant, Rutherford knew that he had to remain abreast of new developments, but his lifelong commitment to simple experiments and apparatus, combined with his ingenuity and perseverance, made for penetrating physics insights. If anything was possible in physics, then Rutherford would try to track it down. This persistence with simple experiments contrasted with Rutherford's breadth of vision as his authority grew. At his insistence, British scientific research flourished, helping to ensure that the UK had no shortage of scientific manpower when the second World War arrived. Co-opted onto many committees, Rutherford was also very influential in stressing the importance of applied research in industry.

Despite Rutherford's successful economical approach to nuclear physics, by the mid-1930s it was becoming increasingly clear that new understanding was going to need major investment. Despite Rutherford's insistence to the contrary, nuclear physics had begun to outstrip what could be done using bench-top experiments alone. The prime example was the cyclotron, invented by Ernest Lawrence at Berkeley, California, in 1930 to take beams of electrically charged particles to higher energies. Rutherford was adamant that the Cavendish Laboratory did not need a cyclotron, and James Chadwick, a long-time collaborator and the discoverer of the neutron (for which he earned the Nobel prize in 1935), resigned and moved to Liverpool, where he would be free to build one.

The right man at the right time disappeared at the wrong time, when on 19 October 1937 Rutherford died after intestinal complications following what should have been a routine operation for an abdominal hernia. He was only 66, and was survived by his one-time research supervisor J J Thomson, who Rutherford had succeeded as Head of the Cavendish Laboratory in 1919. Aged 80, Thomson read the lesson at the memorial service for Rutherford in Cambridge's Trinity College Chapel. As what could be done with bench-top experiments was limited, Rutherford's nuclear physics research reign came to a natural end, but one can only speculate on a possible influential role for Rutherford during the Second World War. His insight had certainly been valuable in the First World War, and his stature outside his own laboratory had grown with time.

However, his resistance to the cyclotron did not help British physics. Just a week before his death, he wrote to Chadwick in Liverpool saying 'I have long since recognized that the cyclotron is a costly toy in this country'. With Rutherford's disappearance, a curtain fell on one act of physics, never to be raised again. In the late 1930s, the decisive experiments on the role of neutrons and nuclear fission took place in Rome, Berlin and Paris. Without Rutherford's tenacity and with Chadwick, the father of the neutron, gone to Liverpool, the Cavendish Laboratory had to take a back seat in nuclear physics. After Sir Edward Appleton assured continuity as head of the Cavendish Laboratory, Rutherford's appointed successor was W L Bragg. Just as the discovery of x-rays had influenced the course of Cambridge research at the close of the nineteenth century, so in the 1930s the Cavendish Laboratory set out on a new road, which was eventually to lead to epic discoveries in molecular biology.

While Rutherford's contributions to science loom head and shoulders above those of his contemporaries, he nevertheless retained a curious detachment to his work. As an anonymous obituary to Rutherford in *The Times* enigmatically remarked 'To speak of devotion to science would be to invite his sarcasm, for he was ever scornful of 'these fellows who take themselves so seriously'.' It was perhaps this detachment which made him so objective.

The other aspect of subatomic physics which undermined Rutherford's position was its theoretical implications. Rutherford displayed remarkable analytical powers; his development of 'Rutherford scattering', clarifying the 1909 discovery of the nucleus, remains a classic. It soon became clear, however, that quantum mechanics, the radical new picture required to understand how Nature operated on a microscopic scale, required a totally new approach. The development of quantum mechanics essentially took place in the German language. It was Max Planck in Berlin who had introduced the quantum idea at the turn of the century. To explain the precision measurements on the distribution of wavelengths emitted by hot objects, Planck supposed that radiation was not a continuous stream, like a river. Instead it was like rainfall, where the energy of the 'drops' depended on the frequency of the radiation.

In 1905, Albert Einstein showed how this idea of radiation as quantum droplets (photons) could explain the photoelectric effect, when light falling on a sensitive surface produces a tiny electric current. Einstein explained how individual photon droplets effectively knock electrons out of atoms. Then for a time it looked as though quantum theory would leave its German-speaking kindergarten. In Denmark, Niels Bohr put forward the idea of fixed quantum orbits for atomic electrons, like the rungs of a ladder. The atomic electrons could not sit anywhere and could only jump from one rung to another, emitting or absorbing radiation as they went. Each particular jump produced its own characteristic spectral line. In France in 1923, Louis de Broglie suggested that, if waves contained quantum droplets which behaved like particles in the photoelectric effect, then the reverse could also happen. Particles should also sometimes look like waves, and de Broglie wrote down the appropriate equation. Bohr's electron orbits could be interpreted as the different ways that the electron waves could be fitted into their closed atomic orbits.

The Professor of Theoretical Physics in Zürich in 1925 was Erwin Schrödinger, an Austrian who had achieved physics fame with his work on statistical mechanics and on the theory of colour vision. The arrival of Schrödinger in Zürich in 1921 at the age of 34 as the distinguished successor to Einstein, Debye and Laue could have been the pinnacle of his career, but his greatest achievement was yet to come. In 1925, he began the academic year with a heavy teaching load. A regular feature on the Zurich physics agenda was the joint seminars organized by the university and the nearby Eidgenössiche Technische Hochschule (ETH). Pieter Debye, at the ETH, suggested that Schrödinger should give a seminar on the new ideas of de Broglie. At the end of the seminar, Debye remarked that as a student of Arnold Sommerfeld in Aachen he had been taught that to handle waves properly needed a wave equation. This obviously impressed Schrödinger who, giving a second seminar a few weeks later, began by saying, 'My colleague Debye suggested that one should have a wave equation; well, I have found one!'

Schrödinger himself did calculations with his new equation, but it took the insight of Max Born at Göttingen to understand the physical significance of the mysterious wavefunction. Meanwhile also at Göttingen a young researcher called Werner Heisenberg was pursuing a different route. Realizing that the quantum world required a radically new approach, Heisenberg developed a picture of 'matrix mechanics' in which mathematical operations modelled what happened in the invisible microscopic world. With Born and another young colleague, Pascual Jordan, he worked out the complete formalism of the new mechanics. In 1926 another young researcher, Wolfgang Pauli, applied the new matrix mechanics to the hydrogen atom and arrived at the same results that Schrödinger had obtained with his wave equation.

In the mid-1920s, Göttingen was to theoretical physics what the Cavendish Laboratory was to the experimental science. Under the guidance of Max Born, Heisenberg and Jordan built a totally new picture of quantum physics. Together with Pauli, these young researchers (Pauli was born in 1900, Heisenberg in the following year, and Jordan in 1902) were the prophets of the new ideas. These youthful minds, uncluttered with ideas whose time had past, thrived on the new quantum concepts. That period was known at Göttingen as the years of 'Knabenphysik' (boy physics) but the boys still needed the guidance and insight

Figure 4: Wolfgang Pauli, born in Austria, spent most of his scientific career in Zürich, Switzerland. Intolerant of sloppy thinking and half-baked ideas, his opinion was widely respected and he became known as 'the conscience of physics'.

of experience. Max Born played a vital role, while Heisenberg went to Copenhagen to work with the mature Niels Bohr. The three *éminences grises* of quantum theory, Bohr, Schrödinger and Born, were of a similar age, born in 1885, 1887 and 1882 respectively. Clearly the interaction between the wisdom of these men and the boy wonders at Göttingen was influential in shaping quantum mechanics. From this mecca, the message spread throughout the world. Spending a year at Bohr's institute in Copenhagen or in one of the major German-speaking universities (Göttingen, Munich, Berlin, Leipzig or Zurich) was mandatory for any aspiring theorist. German was acknowledged as the major language of science and young scientists were advised early in their careers to learn it.

However, there was one notable exception to this German-speaking dominance of the theoretical stage. Paul Adrien Maurice Dirac, the son of a Swiss immigrant, was born in Bristol in 1902. After being pushed by his domineering father towards an engineering career but preferring a more mathematical direction, he began theoretical research at Cambridge in 1923. While experimental physics research at Cambridge was flourishing under Rutherford, the theoretical side was a relative backwater. Ralph Fowler was following as best he could the new developments in Germany and introduced Dirac to this new line of thinking. Working totally alone, the young Dirac immediately went off and carved out his own approach to quantum mechanics, using a formalism which parallelled that of Heisenberg but which displayed more openly the subtle analogies between classical and quantum dynamics. Dirac's work was acknowledged to merge the wave approach of Schrödinger with the matrix picture of Heisenberg.

Then, in 1927, Dirac went one step further. Several quantum mechanics pioneers had realized that Einstein's relativity had to be incorporated into the general scheme to make everything complete. Schrödinger had tried to do this for his original wave equation in 1926, but these and other initial attempts were hampered in two ways. Firstly they neglected to take account of a newly discovered quantity, the 'spin' of the electron, and secondly these initial equations could not reconcile the quadratic equations of relativity with the essentially linear equations of quantum mechanics. Dirac's 'designer equation' took account of electron spin and, using a new scheme of four-dimensional matrices, succeeded in expressing relativity in a suitable form. It was a triumph of mathematical ingenuity, but there was a price to pay. The Dirac electron had four components, two corresponding to the two allowed spin orientations of the ordinary electron, and two others, with opposite electric charge to the electron and with negative energy.

While everybody admired Dirac's new relativistic equation for the electron, these unexpected negative-energy states created much confusion. Dirac himself initially tried to explain the opposite charge solutions as the proton, but this was difficult to accept as the proton is some two thousand times heavier than the electron. After hesitating for four years, in 1931 Dirac finally predicted what his beautiful equation had said all along, that the negative-energy positive-charge states corresponded to a new particle, the positive electron, a 'mirror image' of the familiar subatomic electron.

That same year, some 8000 km from Cambridge, at the California Institute of Technology (Caltech), Pasadena, a young researcher called Carl Anderson was looking at cloud chamber photographs of cosmic rays. Equipped with a powerful magnet, Anderson's cloud chamber showed curved tracks, the curvature reflecting the charge of the particles. Some of Anderson's tracks looked like electrons but appeared to curve the wrong way, as though they were carrying positive instead of negative charge. Anderson's research supervisor, Robert Millikan, assumed that the positive charge tracks had to be protons, but Anderson, and other physicists, seeing the tracks that Millikan displayed on a European tour, thought the tracks were too sparse to be protons. Heavy particles such as protons gave lots of ionization and left characteristic thick tracks which contrasted with the straggly footprints left by electrons. Daring to disagree with his supervisor, Anderson suggested in 1932 that his opposite curvature tracks were a new kind of positive electron. Anderson did not know about Dirac's prediction, and it was only when the news of Anderson's discovery reached Europe that the connection was made.

It is ironic that Dirac made his prediction of a positive electron in the same city as the world's leading subatomic physics laboratory, but that the discovery was made on the far side of another continent. True, the positron, as it became to be called, was quickly confirmed at Cambridge by Patrick Blackett and Giuseppe Occhialini but, had the implications of Dirac's work been appreciated several years earlier by the Cambridge experimenters, then they would surely have beaten Anderson to the post. As it was, the unrepentant Rutherford, speaking at the Solvay physics conference in Brussels in 1933, where the positron discovery was discussed, said, 'It seems to be to a certain degree regrettable that we had a theory of the positive electron before the experiments. ... I would be more pleased if the theory had appeared after the establishment of the experimental facts!'

The prediction and discovery of the positron took place against a backdrop of the Great Depression. The New York stock market crash of October 1929 made US banks call in major international loans, particularly those to finance the massive German reparations payments fixed by the Treaty of Versailles after the First World War. In a domino effect, banks across the world had to close. With no finance available, factories had to shut down and workers were laid off. Unrest favoured political extremism, particularly in Germany, where there were six million unemployed. The Weimar Republic, already staggering under the humiliation of defeat and the subsequent burden of inordinate reparations payments, was balanced on a political knife edge. As the balance tipped, the Nazi party swept to power, feeding on a general fear of communism. In this volatile climate, the Jews, as several times before in their history, became the scapegoats for a general malaise. When Adolf Hitler came to power in 1933, the trickle of antisemitism became a flood, and one of the first laws passed by the new regime was to eliminate Jews and socialists from all government positions, including universities. Max Planck, as Secretary of the Berlin Academy of Sciences and President of the influential Kaiser Wilhelm Research Institute, dared voice his disapproval, saying that the removal of Jews would bleed German science white, but his warning fell on deaf ears.

Over the next few years, a wave of German-speaking scientific talent emigrated. Some of them, like Albert Einstein, could command jobs at prestigious universities. Others were less fortunate. Still more were yet to make their major contributions to science and had to beg. Most of these emigrants went to the UK, where major industrial concerns provided grants for students, and to the USA, where the Rockefeller Foundation was one of those who rose to the challenge. Among those pushing for aid for intellectual refugees was Edward R Murrow, formerly President of the US National Student Federation, and who went on to become a famous media personality. The physics emigrants included Hans Bethe, Max Delbrück, Walter Gordon, James Frank, Otto Frisch, Herbert Fröhlich, George Gamov, Maurice Goldhaber, Walter Heitler, Georg von Hevesy, Alfred Landé, Fritz London, Lise Meitner, John von Neumann, Rudolf Peierls, Frank Simon, Otto Stern, Leo Szilard, Edward Teller, George Uhlenbeck, Victor Weisskopf, Hermann Weyl and Eugene Wigner. But the tide of intellectual refugees covered all fields, literature, art and music as well as science and reached its peak in 1938–9.

In Mussolini's Italy, too, antisemitism took its toll. Italy's leading physicist was Enrico Fermi, who had worked in Göttingen with Born and subsequently became Professor of Physics at Rome. In 1938, Fermi, whose wife was Jewish, was awarded the Nobel prize for his discovery of new radioactive isotopes by neutron irradiation and nuclear reactions initiated by slow neutrons, work which set the stage for the next act in the nuclear play. Fermi only bought a one-way ticket from Rome to Stockholm. From Stockholm he went on to a professorship at New York's Columbia University. Other Italian physicist emigrants of the period included Giuseppe Occhialini, Bruno Pontecorvo, Giulio Racah, Bruno Rossi and Emilio Segrè.

Wolfgang Pauli had moved to Zurich in 1928, but even in neutral Switzerland the proximity of what was happening across the border made him feel uncomfortable, and in 1940 he went to Princeton and stayed there for the duration of the war. In 1945 he was awarded the Nobel prize. Heisenberg remained in Germany but came under attack as a 'white Jew' for his insistence on using relativity, denounced as Jewish propaganda by the Nazis, whose physics figureheads were Philipp Lenard (who won the Nobel prize in 1905) and Johannes Stark (1919 Nobel prize). The award of Nobel prizes to researchers such as Einstein and Schrödinger was seen in some quarters as evidence of a Jewish plot to dominate the Nobel prize scene. Politically acceptable but denuded of talent at a vital time, the German theoretical physics scene stagnated.

The outstanding theoretical problem was the quantum theory of radiation, but with continual physics migration from continental Europe preventing the concentration of effort needed to crack this problem, it remained intransigent. In Cambridge, Dirac made valiant efforts, but the inventor of the relativistic equation of the electron preferred solitude and never established a school. The Dutch physicist Hendrik Kramers also made notable progress, but a complete theory of quantum radiation (what is now known as quantum electrodynamics) had to await the emergence of a new generation of vigorous talent in the immediate post-war years, mainly in the USA.

On 20 May 1899, 38 physicists had gathered at Columbia University in New York City to found the American Physical Society. The journal *Physical Review* had been founded six years earlier and in 1903 came under the wing of the new national society. However, even in the late nineteenth century, lecture tours to US universities had attracted leading Europeans such as Helmholtz, Rayleigh and Kelvin. Especially for British physicists, where there was no language barrier, trips across the Atlantic beckoned. A major US achievement came when Arthur Holly Compton in St Louis discovered that x-ray wavelengths increased after scattering by electrons, showing that they lose energy. This dramatic illustration that electromagnetic radiation has definite particle characteristics earned Compton, by then at Chicago, the 1927 Nobel prize, the third American to receive the award in a total of 32 recipients. Compton's revelation also predated the 1929 discovery of electron diffraction, usually cited as the classic demonstration that particles can also behave as waves.

A trip to one of the major European centres was mandatory for aspiring US physicists. Compton had worked at Cambridge in 1919, and in 1926 there were more than 20 US visitors at Göttingen alone. Another visitor to Cambridge was one J Robert Oppenheimer, who initially wanted to work on experiments in Rutherford's Cavendish Laboratory but turned to theory instead, where he was a contemporary of Dirac. Oppenheimer was clearly emotionally disturbed during those few years, prone to depression and extreme bursts of temper. After Cambridge, Oppenheimer moved first to Leiden, then to the mecca at Göttingen, where he clearly impressed Born, and finally to Pauli in Zurich, who remarked that Oppenheimer spent more time undergoing psychoanalysis than doing

Figure 5: After studying physics in Europe, J Robert Oppenheimer guided a new generation of US theoretical physicists before becoming the scientific director of the wartime Manhattan Project which developed the atomic bomb.

physics. (Pauli himself, worried by the inadequacies of his own immense powers of rational thought when faced with everyday problems, was to spend many hours with the pioneer Swiss psychiatrist Carl Gustav Jung, who found Pauli a fascinating subject.) However difficult it had been for Oppenheimer personally, his exposure to the furnace of European theoretical work prepared him for his return to the USA in 1929, where he took up parallel appointments at Berkeley and Pasadena and became the US prime exponent of quantum theory. During this time his students included Gregory Breit, Edward Condon, Willis Lamb, John Slater, Philip Morrison and John Van Vleck, while Berkeley became a haven for young postdoctoral fellows. Although Oppenheimer went on to achieve another kind of fame as the scientific director of the Manhattan Project to develop the atomic bomb at Los Alamos during the Second World War, Abraham Pais in his book *Inward Bound* says, of Oppenheimer, 'I do ... believe that none of his contributions deserves higher praise and will longer endure than what he did for the growth of American physics in the decade before the Second World War.'

Also at Berkeley was Ernest Orlando Lawrence, who in 1929 turned his attention to the acceleration of charged particles by electric fields. As a boy in South Dakota, Lawrence had tinkered with radios and understood the importance of electrodes. To pierce the nucleus, Rutherford had long realized that alpha and beta particles from ordinary radioactive decay were insufficient and saw the need to accelerate particles to a high energy. The original late

Figure 6: *In 1929 at the University of California campus at Berkeley, near San Francisco, Ernest Orlando Lawrence turned his attention to the acceleration of charged particles by electric fields and developed the cyclotron.*

nineteenth-century cathode-ray experiments which had discovered x-rays and the electron needed high voltages and a fair vacuum, supplied by an induction coil and improved pumps, but for nuclear work these voltages were insufficient. Accelerating particles also needed a good high-vacuum tube. In the USA, efforts to accelerate particles got under way at the Department of Terrestrial Magnetism of the Carnegie Institution in Washington with Gregory Breit and Merle Tuve. In 1931 at Princeton, Robert Van de Graaff perfected his famous belt-driven electrostatic generator.

At Berkeley, Lawrence, who was an old schoolfriend of Tuve, stumbled across an 1927 article in *Archiv für Elektrotechnik* by Rolf Wideröe, a Norwegian engineer working in Aachen. Wideröe had accelerated charged particles by passing them through an increasingly long series of cylinders with appropriate polarities, the forerunner of the linear accelerator, or linac. Lawrence could not understand the German but, by looking at the accompanying diagrams, soon got the gist of the idea. To produce a high-energy beam, the Wideröe apparatus would be far too long to fit into a conventional laboratory. Lawrence's brainwave was to make the particles rotate in a magnetic field and then to apply Wideröe's idea. The particles would then spiral round and round, receiving a kick in energy each time that they passed through the voltage gap of an oscillator. Instead of a tube a block long, the idea could be made to work on a table top, and the development of radio technology meant that the necessary high frequency oscillators were available. During 1930, Lawrence tinkered with the new 'cyclotron' idea, and his successful demonstration the following year with Stanley Livingston put Berkeley on the physics map. However, with money difficult to come by at the peak of the depression, Lawrence had to beg for equipment and funds, promising that his new techniques would bring improved medical therapy and fresh sources of energy.

Meanwhile, in Rutherford's laboratory, John Cockcroft and Ernest Walton succeeded for the first time in combining a voltage sufficiently high to break nuclei with a high-vacuum tube. In the world's first high energy physics experiment in 1932, they smashed lithium nuclei with artificially accelerated protons. Lawrence, who managed to intercept this piece of scientific news on his honeymoon, sent a telegram to Berkeley instructing his laboratory to procure some lithium from the nearby chemistry department and to confirm the discovery of Cockcroft and Walton. The following year, Lawrence was invited to the Solvay conference in Brussels where he described a series of cyclotron experiments on the scattering of deuterons by different nuclei. The deuteron, the nucleus of heavy hydrogen, had been discovered by Harold Urey at the National Bureau of Standards in Washington only the previous year. Quizzed at the Solvay meeting by the European high priests on what they saw as inexplicable results, with similar behaviour for a wide range of targets, Lawrence returned to Berkeley to find his results had been spoiled by deuterons sticking to the targets, so that he had been misled by the scattering of deuterons from deuterium impurities implanted by the beams. Lawrence kicked himself and castigated his staff for doing sloppy experiments.

Then, in February 1934, Marie Curie's daughter, Irène, and her husband, Frédéric Joliot, working at the Institut de Radium in Paris, announced the discovery of artificial beta-radioactive nuclei, formed when light metals such as aluminium were irradiated with alpha particles. Again Lawrence had missed the boat. His cyclotron had been producing such isotopes for the past three years but, because the cyclotron and the Geiger counter which monitored what the cyclotron was producing were both powered through the same switch, when the cyclotron was switched off, so was the counter, and any residual radioactivity was never measured. Rewiring the apparatus, they were again able to confirm a major European result. Undeterred by these initial setbacks, Lawrence concentrated on his major objective, building larger and larger cyclotrons. The energy of these machines are measured in electronvolts (abbreviated to eV), 1 eV being the energy gained by an electron when it passes across a potential difference of 1 V. High energies are expressed in terms of megaelectronvolts (MeV), gigaelectronvolts (GeV) or teraelectronvolts (TeV). From his initial 4 inch demonstration model with its 80 000 eV, Lawrence had progressed by 1939 to a 60 inch machine producing 19 MeV.

For the next quarter of a century, the USA was to hold a monopoly on high-energy machines. As European laboratories saw the value of having a cyclotron, Lawrence's expertise was invaluable in getting machines off the ground at Liverpool (Chadwick), Paris (Joliot) and Copenhagen (Bohr). While Lawrence's cyclotrons missed several major physics discoveries, they did open one valuable new applications area. After having realized that his cyclotrons could synthesize radioisotopes, Lawrence, with his brother John, produced radioactive tracers to diagnose and treat diseases, the first use of artificial isotopes in nuclear medicine.

While Lawrence was pushing cyclotrons at Berkeley, to the south of San Francisco, at Stanford University, Palo Alto, William Hansen was also looking at new ways to accelerate particles. Using a magnet to make electrons shuttle to and fro inside a cavity, Hansen developed a device called the rhumbatron. Russell Varian, a student of Hansen, had a brother called Sigurd, a former pilot for Pan-Am, who had extensive knowledge of the air routes in central and South America. Sigurd, worried that something like the Spanish Civil War could spread to Latin America and hence to the USA, discussed with his brother the idea of a radio 'searchlight' to scan the sky for aircraft. Invited to move their laboratory to Stanford, the Varian brothers realized that Hansen's new rhumbatron was just what they were looking for and developed from it the klystron, which went on to play a major role in microwave work.

In Europe, the Joliot–Curie discovery of induced radioactivity which Lawrence had overlooked had made Paris jubilant. After missing the neutron and the positron, France, where Becquerel had discovered radioactivity in 1896 and where the Curies had made scientific history with the identification of radio-active sources, was back in the forefront of physics. The 1934 breakthrough by her daughter and son-in-law gave Marie Curie a final satisfaction before she was to die of leukaemia several months later.

In Rome, Enrico Fermi pounced on the idea of using neutrons to bombard nuclei. While alpha particles have to fight against the positive charge of the nucleus, neutrons, being electrically neutral, are assured of an easier passage. Within four months of the Joliot–Curie discovery, Fermi had discovered some 40 new radioactive nuclei. Later that year, two of Fermi's collaborators, Bruno Pontecorvo and Edoardo Amaldi, discovered that more artificial radioactivity was produced when the apparatus stood on a wooden table rather than a metal plate. Reporting their finding to Fermi, he suggested surrounding the whole experiment by a block of paraffin wax. Repeated collisions in the hydrogen-rich wax slowed down the neutrons, making them more effective at inducing radioactivity.

As the news of the Rome discoveries spread across Europe, research focused on these new effects. At Bohr's institute in Copenhagen, the Italian papers were translated by Otto Frisch, an Austrian refugee. In Berlin's Kaiser Wilhelm Institute, relatively unscathed by the 1933 race laws, the news aroused the interest of two radiochemists, Otto Hahn and Lise Meitner, who was Otto Frisch's aunt. The reactionary director of the institute, Emil Fischer, unused to the idea of women scientists, had first made Meitner work in the basement of the building, but thirty years later her persistence and undoubted ability had established her reputation. One of the elements that Fermi had irradiated with neutrons was uranium, and he assumed that the isotopes being produced were close relatives of uranium. As radiochemists, Hahn and Meitner, who were joined by Fritz Strassmann, were curious to know exactly what isotopes were produced by bombarding uranium by neutrons. So was Irène Joliot-Curie in Paris but, with Frédéric Joliot distracted by fund raising and by politics, the answer eluded her.

When Austria was annexed by Germany in 1938, all Austrian citizens became German and were subject to the same race laws as Germans. Lisa Meitner had no option other than to leave Germany. Although refused an exit visa, she crossed illegally into Holland and went on to Stockholm. Continuing on without Meitner, just before Christmas of that year Hahn and Strassmann finally solved the riddle of the uranium neutron irradiation. The difficulty had been compounded by the need for an unusual range of expertise—physics skills to understand the nuclear processes, and chemistry skills to identify the elements produced. Rather than producing a large kindred nucleus, the giant uranium nucleus could not digest the extra neutron and instead split into two roughly equal nuclei, bromine and lanthanum. No such radioactive process had ever been seen before. In a letter to Lisa Meitner, Hahn described it as a 'horrifying conclusion, contradicting all previous experience'. Otto Frisch travelled from Copenhagen to Stockholm to spend Christmas with Meitner and found his aunt puzzling over Hahn's letter. Over the holiday, they realized that, in splitting in two more stable nuclei, the uranium nucleus was shedding surplus energy.

While Hahn had described his and Strassmann's discovery as a 'horrifying conclusion', he was not yet aware of the implications. Their paper was rushed into print in *Naturwissenschaften* but, because of the end-of-year holidays did not appear until 6 January 1939, which was how most other European scientists learned the news. However, Frisch and Meitner had been forewarned by Hahn's letter. In a letter to the journal *Nature*, published on 16 January 1939, they described 'a new type of nuclear reaction', which they called 'fission', comparing the unstable uranium nucleus with a dividing biological cell. Returning to Copenhagen before these papers were published, Frisch tipped off Bohr, who was preparing for a trip to the USA. Who better than Niels Bohr to inform US physicists of the discovery of fission? Fermi, who probably would have discovered fission himself had he had the chemistry support to identify the produced elements, was already at Columbia University, New York. With European science benumbed by the imminence of war, Bohr's news of nuclear fission fell on fertile ground in the USA.

THE STORM BREAKS

At the British Association for the Advancement of Science meeting in 1933, before nuclear fission by neutrons had been discovered, Rutherford went on record as saying that talk of energy from nuclear processes was 'moonshine'— not the most prophetic of words for a meeting to speculate on future directions in science. One who did not share Rutherford's 1933 viewpoint was Leo Szilárd, a Hungarian refugee from the Nazi race laws. In a British patent application dated 12 March 1934, he predicted that uranium could supply energy. In 1935, the husband-and-wife team of Frédéric Joliot and Irène Curie received the Nobel chemistry prize for their work on artificial radioactivity, 32 years after the previous Curie husband-and-wife team had shared the Nobel physics prize with Henri Becquerel for their pioneer work on radioactivity. In his Nobel lecture, Frédéric Joliot surmised that 'chain reactions' could be produced by nuclear transmutations, with an 'enormous liberation' of energy, once they could become self-sustaining.

With the discovery of uranium fission by neutrons in 1938, the suspicion grew that this reaction could be the source of new neutrons to catalyse more fissions and to make the reaction self-sustaining. Niels Bohr, while still on his early 1939 US tour, suspected that such a chain reaction would more likely occur with the rare isotope uranium-235, a 0.7% 'impurity' in natural uranium, rather than ordinary uranium-238. Nuclear particles, like neutrons, like to pair together. Uranium-235, with an odd number of neutrons, would readily form the unstable fissionable nucleus uranium-236. Uranium-238, on the other hand, would tend to mop up naturally occurring neutrons rather than to become fissionable.

Even before the move away from nuclear physics at the Cavendish Laboratory in Cambridge after Rutherford's death in 1937, Frédéric Joliot's laboratory at the Collège de France was already attracting researchers from all over Europe. The uranium neutron chain reaction predicted by Joliot in 1935 was first detected there by two immigrant researchers, Lew Kowarski, who was born in St Petersburg, and who had made his way to France via Poland and Belgium, and the Austrian Hans von Halban, who had worked with Frisch in Copenhagen. Pumping neutrons into the centre of a tank of uranium salt solution, they found that the neutron distribution towards the rim of the tank did not fall off as quickly as with other substances. The same discovery was made at Columbia University by Fermi working with Szilárd, who had by then moved to the USA from Britain. While Kowarski opted to airmail the news of the discovery for publication in the widely read British journal *Nature*, Szilárd, mindful of his 1934 ideas on nuclear energy, advised Fermi that they should keep quiet. The Second World War was only months away and Szilárd did not want German science to benefit. In fact the Hungarian had already written to Joliot advising him not to publish but was overruled by French enthusiasm. With scientists vying for fame, the idea of a voluntary ban on fission publications fell on infertile ground, and the results of Fermi and Szilárd, too, were published shortly afterwards. That year more than a hundred scientific papers on fission appeared. Each new development was eagerly snapped up in the USA, France, Germany, the Soviet Union and Japan.

A vital figure was the number of neutrons produced in each fission reaction. Some neutrons would always be lost, so the target figure to ensure a self-reproducing reaction was about two neutrons per fission. Soon the Paris experiments were reporting 3.5! However, an error was found, bringing this figure to 2.6. Another vital concept was reintroduced by Francis Perrin in Paris in 1939, although Szilárd had noted it in his 1934 patent. It was the idea of a critical size. If the piece of uranium is too small, too many neutrons will leak out of the sides before they get a chance to interact with other uranium nuclei. Uranium sources were immediately at a premium, and Edgar Sengier, President of the Union Minière du Haut-Katanga in the Belgian Congo, suddenly found one of his commodities very much in demand, with high-level contacts from many countries. At that time, much of the demand for uranium ore was because of its tiny radium content, with the bulk of the 'low-grade' uranium content traditionally going unwanted. This uranium 'waste' was soon snapped up.

Another vital parameter was the speed of the neutrons producing the chain reaction. Fast neutrons tend to blunder around and to lose energy in all sorts of ways, while slow neutrons have a better chance of catalysing fission in uranium-235. Two scenarios opened up. In one, the neutron velocities could be optimized by a suitable 'moderator' so that even the tiny fraction of the 235 isotope in natural uranium might be enough to support a self-sustaining but nevertheless controlled reaction. In the early 1940s, many experiments using different moderators and reactor geometries strived to attain criticality.

The second scenario depended on having more uranium-235 than occurs naturally—'enriched uranium'. If sufficient uranium-235 were available, then the fission level produced by fast neutrons could make the reaction self-sustaining. Because of the speed of the neutrons, they would multiply much more quickly. The nuclear reaction could get out of control and cause an explosion. Estimating the amount of energy released, the fission scientists realized that such an explosion could make a bomb based on a conventional chemical explosive seem puny by comparison—it would be a 'superbomb' (although not as super as what was to come later with the fusion bomb).

Although Bohr, working with John Wheeler, had predicted broad outlines, the detailed properties of uranium-235 were totally unknown, and predictions of the critical mass of uranium needed to make a superbomb ranged from less than a kilogram to 40 tonnes, if it would work at all.

In the summer of 1939 in the USA, the Hungarian immigrants Leo Szilárd and Eugene Wigner drafted a letter to President Roosevelt pointing out the potential of a fission bomb. To increase its impact, rather than signing it themselves, they gave their draft to Albert Einstein and persuaded him to write to the US President. Duly impressed, the US President called for action, and in October 1939 a US Advisory Committee was set up under Lyman J Briggs, Director of the US National Bureau of Standards. In the UK, the French work on neutrons attracted the interest of George Thomson, son of J J Thomson, at London's Imperial College, who spread the message to his colleagues. At government level, opinion was sceptical, but the potential power of any fission bomb counterbalanced the scepticism and uranium development was place in the hands of the Air Ministry, who purchased a ton of uranium oxide, to be put at the disposal of Mark Oliphant at Birmingham, who had previously worked under Rutherford at Cambridge. Oliphant, born in Australia in 1901, had a particular empathy with the New Zealander Rutherford. Oliphant's 1934 discovery of tritium, the third isotope of hydrogen, was probably the last in the long series of major nuclear discoveries in which Rutherford participated. In 1937, Oliphant left Cambridge to set up what would become a dynamic new school in Birmingham, accentuating the depletion of nuclear physics expertise at Cambridge.

Although the powers-that-be had been alerted to the possibility of a fission bomb, even among scientists the idea seemed too far fetched, and in 1939, in France, the USA, Britain and Germany, the idea of a fission bomb was put on the back burner. The immediate goal was instead the development of a self-sustaining or critical fission reactor, or 'pile', as Fermi, with his new command of English, chose to call it. Various arrangements were tried, dissolving uranium compounds in water, hanging containers of uranium oxide in water or interspersing uranium oxide containers in paraffin wax, each of which had various geometrical alternatives. The French achieved encouraging results with a 50 cm sphere of wet uranium oxide in a tank of water.

However, there was another possible route to exploit fission, which had been pointed out in an incisive paper by Niels Bohr and his former associate John Wheeler. Developing Bohr's earlier ideas that uranium-238 was a damp squib as far as fission was concerned, they pointed out that, while the rare uranium-235 should be the optimal fission medium, there were also fission possibilities with an artificial nucleus of atomic weight 239. The publication of this influential paper coincided with the German invasion of Poland and the outbreak of the Second World War in Europe.

The first to take note were the Germans. Scientific manpower had been mobilized, including Otto Hahn, Walther Bothe, Hans Geiger and Carl von Weizsäcker, as well as Werner Heisenberg, who realized different approaches were required for a bomb and for a self-sustaining reactor. One reactor possibility was to use heavy water as 'moderator' to slow down neutrons. In heavy water, the hydrogen atoms of normal water are replaced by atoms of deuterium, whose nuclei contain a neutron as well as the single proton of ordinary hydrogen. With the extra neutron, the deuteron nuclei of heavy hydrogen should be more effective for slowing down neutrons than are the single protons of ordinary hydrogen (which can also absorb them and stifle a chain reaction). Heavy hydrogen occurs naturally as a one part in 5000. Almost the entire world stock of heavy water was in Norway, where, because of its slightly different boiling point to ordinary water, it accumulated as a by-product in the manufacture of ammonia for fertilizers, and IG Farben representatives placed a German government order for the entire stock, plus a request for production to be stepped up. The French, who were also planning to build a reactor with heavy water as moderator, got wind of the German order. Realizing that the Germans could only want it because they too were trying to build a nuclear reactor, they astutely managed to outwit the Germans, transferring all 185 kg to Paris via Britain, despite concerted German attempts to intercept it.

When Germany invaded France in 1940, the heavy water, together with Lew Kowarski and Hans von Halban, was hurriedly sent to safety in Britain, but the Germans, who took over the Paris and Copenhagen laboratories and their cyclotrons, were in a good position to capitalize on ongoing research. Wolfgang Gentner, who had earlier worked at Joliot's laboratory, was despatched to Paris to supervise the completion of Joliot's cyclotron. Although Gentner was on an official government mission, his long-standing friendship with Joliot prevented the Paris cyclotron from contributing significantly to weapons research.

In the summer of 1940 the fission climate changed when a succinct report by Otto Frisch and Rudolf Peierls refocused the British Government's attention on the possibility of a bomb. Frisch, who had been working in Bohr's institute in Copenhagen, had been visiting the UK when war broke out and preferred to stay at Birmingham with Peierls. According to Frisch and Peierls, a few kilograms of uranium-235 could make a fission bomb, releasing a titanic explosion whose effect would be complemented by the release of radiation levels which could endanger life. However they pointed out that such a task would not be easy. The extraction of this amount of uranium-235 (by gaseous diffusion) from parent uranium would require a vast investment. In the wake of the Frisch–Peierls memorandum, a special committee (named MAUD after Bohr's *au pair* but officially designated as Military Applications of Uranium Detonation) set up university-based groups including those of Otto Frisch, Rudolf Peierls, John Cockcroft, James Chadwick and Francis Simon; Imperial Chemical Industries and Metropolitan–Vickers were also called in. When compared with this new mobilization, the arrival of Kowarski and von Halban in Britain with their heavy water was a low-key affair, and their request to push ahead with reactor development was counter to official policy. Kowarski and von Halban were accommodated at Cambridge and subsequently moved to the safety of Montreal.

At Ernest Lawrence's Berkeley laboratory, Edwin McMillan and Philip H Abelson discovered that uranium fission was not the end of the nuclear road. Uranium, when bombarded by neutrons produced via a Berkeley cyclotron, could also produce a new heavy nucleus, number 93 in the periodic table of elements, highly unstable and not found naturally. It was the first transuranic nucleus, which they called neptunium (Neptune being the planet adjacent to Uranus). In its radioactive decay, one neptunium neutron beta-decayed into a proton, in principle producing nucleus number 94. While this nucleus was not initially detected, the Berkeley scientists correctly assumed that this was because nucleus 94, which, continuing the planetary theme, they had already decided to call plutonium, was more stable than neptunium. From its proton–neutron content, it was easy to see that plutonium should be readily fissionable, as Bohr and Wheeler had predicted.

With the paper of McMillan and Abelson openly published, scientists in Britain, Germany and Russia saw the potential of nucleus 94. Because it was a different chemical element from uranium, plutonium could be separated using relatively straightforward chemistry, instead of the tedious mechanical diffusion or centrifugal methods needed to separate the chemically indistinguishable uranium-235 from uranium-238. However, the British MAUD committee had already progressed with plans for a uranium-235 bomb and in April 1941 published a report pointing out the feasibility of a 10 kg bomb of uranium-235, obtained by filtering uranium hexafluoride through many layers of fine mesh. The plant required to achieve this would be very large. At this point, the academic MAUD committee was superseded by the 'Tube Alloys' team, which included industry representatives who knew and understood nothing of nuclear physics but were appointed to look after the purely industrial side of the project.

In the USA, work pushed ahead on several fronts. Fermi and Szilárd, striving to build their self-sustaining reactor, had discovered the effectiveness of graphite as a neutron moderator. In January 1941, Glenn Seaborg at Berkeley detected plutonium in the products of a cyclotron experiment. Within months, the Berkeley physicists demonstrated that plutonium was fissionable. Such discoveries were still being openly published in the scientific literature. However a cyclotron could not be used to mass-produce plutonium, although the first microgram samples were obtained this way. Large-scale production of plutonium needed a reactor rather than a cyclotron, backed up by radiochemical separation techniques. For the uranium-235 route, Jesse W Beams at the University of Virginia was proposing a centrifuge method, while Lawrence at Berkeley and John Dunning at Columbia advocated magnetic separation.

With so much at stake and with so much unchannelled enthusiasm, uncoordinated proposals sprang up all over the place. Despite Roosevelt's earlier Uranium Committee under Briggs, nobody was masterminding the US nuclear effort. The man who stepped into the breach was Vannevar Bush, a patriot and an imaginative electrical engineer who at the Massachusetts Institute of Technology (MIT) had built an early form of computer and had then moved on to become President of the Carnegie Institute. Aware of the threat of war, Bush got Roosevelt to appoint him as Chairman of a new National Defense Research Council, with James B Conant, an influential Harvard chemist, as his assistant. The Briggs committee, disguised as 'S-1', was brought under this new umbrella, while Arthur Compton chaired a National Academy of Sciences review committee.

There was already a feeling that things were really beginning to get moving when, in the summer of 1941, the MAUD report arrived in the USA. This was the spark that set Vannevar Bush's plans alight. Conant persuaded Lawrence to come aboard and, by the time of the Japanese attack on Pearl Harbor which brought the USA into the war, Lawrence's cyclotrons, using an ingenious technique invented by Lawrence to improve their separation potential, had shown that they could separate uranium isotopes. Bush appointed two other American Nobel prize winners, Compton and Urey, as well as Lawrence, as his lieutenants. While Lawrence and Urey were working on uranium separation, which they knew something about, Compton's responsibilities included plutonium production, an almost unexplored field.

Realizing that much construction and logistics work would be needed, Bush and Conant turned to the US Army Corps of Engineers, and for a while the discipline of the Army and the intellectual enthusiasm of the scientists were difficult to reconcile. Realizing that a firm hand was needed, the Army appointed Brigadier General Leslie R Groves, who had supervised the building of the Pentagon in Washington, to take command of the whole fission weapons programme, under the code name 'Manhattan Engineering District', customarily abbreviated to the 'Manhattan Project'. One of his initial decisions was to bring down a curtain of secrecy over the whole operation. This dismayed the British, who had fed their MAUD information to the USA. The sudden interruption in the publication of scientific results was noticed elsewhere. In Russia, Georgi Nikolaevich Flerov, an influential young researcher who had made his mark on fission physics and understood the implications of nuclear energy, spotted that the steady stream of fission-related papers in the *Physical Review* had suddenly dried up. Realizing that the work could not have been stopped, he concluded that the USA had embarked on a military programme towards a bomb and warned Moscow. The warning was heeded, and the Soviet nuclear weapons programme began.

With resources spread all over the country, Compton established in Chicago a central unit, code-named the 'Metallurgical Laboratory', customarily abbreviated to the 'Met Lab'. A site was readied as quickly as possible in the Argonne forest south of Chicago, but in the meantime Fermi's team began to build a reactor in an old squash court under the Stagg Field Chicago University football stadium. There, on 2 December 1942, at 3.30 pm, the uranium–graphite lattice went critical. Compton phoned the news to Conant with the coded message, 'You'll be interested to know that the Italian navigator has just landed in the New World.' Three months later, the prototype pile, its initial work done, was dismantled and re-erected at the new Argonne laboratory. Managed by the Du Pont organization, big reactors to synthesize plutonium were built at Hanford, Washington, and Oak Ridge, Tennessee.

Laboratory techniques could only produce the nuclear explosives in microgram quantities. To take on the gigantic challenge of producing uranium-235 and plutonium in kilogram quantities meant that the manufacturing process had to be multiplied a billionfold. This new nuclear technology required the marshalling of the largest single endeavour that the planet had ever seen. The full might of US industry was tapped, with Du Pont, Union Carbide, General Electric, Westinghouse and Chrysler all involved. At Oak Ridge, the plant built to separate uranium-235 using 'calutrons', a modified form of Lawrence's cyclotron, had 22 000 operators, all beavering away in complete ignorance of what exactly they were doing. The calutron electromagnets called for 86 000 tons of high quality electrical conductor, supplied as silver on loan from the US Treasury. For plutonium, which by 1943 was still only available in milligram quantities, the challenge was equally prodigious. The first half-gram sample was obtained in March 1944 but, by early 1945, the vast complexes at Oak Ridge and Hanford had been marshalled to produce the kilogram quantities needed for actual weapons. In parallel, scientists had to explore the properties of the new metal so that it could be shaped into bomb components.

The British effort, which by now had been transferred to Canada to be out of the range of the Luftwaffe and had been put under the direction of John Cockcroft, eventually closely collaborated with the US programme, and several members of the British team appeared in the USA. The Germans, with Heisenberg, almost managed to construct a working reactor but suffered from having no national master plan. During the preparations for this work, Heisenberg and von Weizsäcker went to see Niels Bohr, who was still in Copenhagen. One version of the story says that their aim was to convince Bohr that the German fission bomb effort was leading nowhere, so that the message could be passed to the

Figure 7: *After introducing a new quantum picture of the atom and presiding over the development of quantum mechanics, in 1939 Niels Bohr pointed out new implications for nuclear fission. In 1943 he was smuggled from Stockholm to Britain before going on to Los Alamos, where his arrival was a major boost for the Manhattan Project.*

Allies, who would then reduce their atomic weapons effort. Another version says they wanted Bohr's advice. In any case, Bohr felt exposed and vulnerable. This vulnerability increased in August 1943 when the Germans declared martial law in occupied Denmark. The British, as well as Bohr himself, feared that the famous scientist would be deported to Germany. Secret messages were smuggled through urging him to escape, and Bohr fled to neutral Sweden in a fishing boat. From there, a Mosquito bomber took him to the UK. The only place for Bohr was in the aircraft's bomb bay. Flying high over Norway to escape detection, the pilot instructed Bohr over the intercom to switch on the oxygen supply. Bohr did not hear and fell unconscious but revived as the aircraft descended as it approached home. Thus another eminent emigrant scientist joined the Allied war effort.

In the USA, a seething mass of interrelated problems had to be solved. However, the Manhattan Project focused on a single major goal—the development and construction of the actual bombs. This came under Compton's charge but, as scientific overseer, Lawrence recommended to Compton his Berkeley colleague, J Robert Oppenheimer. In May 1942, Oppenheimer was transferred to the Met Lab but, as time went on, found it difficult to coordinate the activities of scientists working at widely separated institutes. Communication and coordination could waste valuable time, and Oppenheimer suggested to Groves the idea of a special dedicated laboratory, where some fifty scientists could work together on their common task. The Los Alamos, New Mexico, site was chosen in November 1942. Although Oppenheimer, a pure research scientist, had not yet demonstrated leadership ability, Groves appointed him as the director of the

Figure 8: *Prominent theorists Hans Bethe (left) and J Robert Oppenheimer played major roles in the wartime Manhattan Project. After the war, Bethe went to Cornell, and Oppenheimer became Director of Princeton's Institute for Advanced Study. After being humbled by accusations of being a spy, Oppenheimer died in 1967.*

new laboratory. Oppenheimer was to succeed brilliantly and went on to epitomize a new breed of specialist, the scientist–manager. Despite a high level of patriotic and political motivation, the scientists at Los Alamos frequently felt uneasy about their huge responsibility, and the development of a superbomb weighed heavily on their consciences. However the sudden arrival of the highly committed Niels Bohr (code-named Nicholas Baker at Los Alamos) in their midst lessened their self-doubt.

With the realization that science was going to play a major role in the forthcoming conflict, the US universities had been scoured for talent. By the time of Pearl Harbor, some 1700 physicists had been recruited for defence work of one kind or another, including the Manhattan Project. Among the mobilized physicists was Robert Rathbun Wilson, who had worked with Lawrence at Berkeley before moving to Princeton, and who received a telegram from Lawrence explaining what needed to be done. Born in Frontier, Wyoming, in 1914, Wilson learnt toolmaking and blacksmithing as a boy. His intellectual ambitions took him to Berkeley, where initially he had wanted to study philosophy but turned instead towards the more tangible attractions of physics. At Princeton, Wilson went to work on his own idea to use cyclotron techniques to separate uranium-235 from uranium-238. In the early 1940s, Princeton boasted an impressive array of scientific talent. As well as the legendary Albert Einstein, there was the refugee Eugene Wigner, who in Germany had made major contributions to quantum mechanics, and John A Wheeler, who, with his former collaborator Niels Bohr, had predicted that uranium-235 held the key to fission. Wolfgang Pauli from Zurich was also a visitor, although Pauli was in no way connected with the subsequent war effort.

Among the Princeton students was a young graduate called Richard Feynman. At Princeton, he had expected to work with Wigner but found himself allotted to the much younger Wheeler instead. It turned out to a fruitful match, even if the extrovert New York student and the urbane Princeton physicist seemed an odd contrast. As a teenage boy in New York's Atlantic Coast suburb of Far Rockaway, Feynman had preferred exotic mathematical equations to collecting postage stamps. Feynman was a mathematical wizard who could visualize the behaviour of complex functions in his mind, watching a series expand or seeing an equation generate a curve. Like a jobbing mechanic carrying his own set of spare parts for any eventuality, Feynman was an avid collector of mathematical devices and accumulated an enormous inventory of analytical gadgets.

At high school, this mathematical ability had put him into his school's team in a hotly contested Algebra League. There are two ways of solving mathematics problems. One, the conventional way, involves cranking the handle until eventually the answer comes out. For a tough problem, this demands much perseverance. The other way of solving the problem, the smart way, usually starts with the word 'consider', and the answer just falls out. However, finding what has to be 'considered' in the first place requires unusual ability. In these school contests, Feynman would frequently sit back while all around him frantically scribbled away. Then he would slowly lean forward, pick up his pencil, write a number on his answer sheet, draw a ring round it, put his pencil down and then smugly sit back again.

In his final undergraduate year at MIT, Feynman entered the Putnam

Figure 9: *The exuberant genius of Richard Feynman (seen here giving a talk at CERN in 1970) first flowered at Los Alamos during the Second World War.*

mathematics competition for a scholarship at Harvard, where the questions were so hard that for normal mortals scoring any marks at all was difficult. Feynman walked out of the exam before the allotted time was up, with a score way ahead of the next best candidates, but the genius had already decided to go to Princeton. His mathematical ability and flashes of brilliance inevitably impressed more and more people. With Wheeler, he developed a new picture of interacting electrons which was to prove a useful testbed for a new kind of quantum theory but, because of the imminence of war, Princeton physics began to group around a small number of war-related projects, including Wilson's cyclotron scheme for separating uranium isotopes, which Feynman joined. Lawrence, trying to avoid overlapping activities, closed down Wilson's scheme at Princeton, so that Wilson and Feynman were temporarily at a loose end. Before they were to embark on their journey to Los Alamos, Feynman made a brief trip to Chicago for a meeting with the Met Lab people.

To be head of theory at Los Alamos, Oppenheimer appointed Hans Bethe, a gifted German physicist who had come to Cornell in 1935 via Britain after being expelled by the 1933 German race laws. Bethe's deputy at Los Alamos was the Austrian Victor Weisskopf, who had worked at Göttingen and Berlin before going to Zurich to be Pauli's assistant. The volatile Richard Feynman was assigned to Bethe's command. With little instinctive respect for authority, Feynman's genius could be difficult to control, but his ebullience met its match in the stoic Bethe. Feynman was the ideal foil to Bethe, and Bethe took note.

The problems being attacked at Los Alamos needed accurate numerical answers. The assembly of a fission bomb had to be carefully controlled so that it remained subcritical, with its nuclear explosive, uranium-235 or plutonium, spread out. The idea was that a conventional explosive detonator coating outside the fission components would compress the fission fuel together in one lump, and the chain reaction would take over. In principle it was easy, but in practice the exact size of every component had to be carefully calculated. With no electronic computers yet available, the theory people were continually kept busy solving equations and supplying numbers on demand. Also on hand was John von Neumann, the Hungarian mathematics genius who went on to play an important role in the development of computing techniques.

The fission bomb was the ultimate offensive weapon of the time. The other new technology pushed during the Second World War was radar, a revolutionary means of defence. In an unusually prescient development in a country so often hampered by muddled thinking, in 1934–5 the British Air Ministry had set up a scientific committee to examine air defence. Chaired by Henry Tizard, formerly Secretary of the Department of Scientific and Industrial Research, the committee also included the following: Air Ministry's Director of Scientific Research, H E Wimperis; A P Rowe, who was Wimperis' assistant; Patrick Blackett, then at Birkbeck College, London; A V Hill, who had won the Nobel prize for physiology. Among the suggestions that the committee attracted was one by Robert Watson-Watt for detecting approaching aircraft by the reflection of pulsed radio signals. Watson-Watt had initially been asked to look at the possibility of using radio waves as an offensive weapon but concluded that not enough energy was delivered. In any case, aircrew would be safe inside metal aircraft. First called radiolocation or Range and Direction Finding (RDF), Watson-Watt's technique used radio wavelengths of about 1.5 m. Such equipment was clumsy, especially for airborne use, and had limited accuracy. For use in aircraft, smaller aerials and smaller wavelengths were vital.

The major breakthrough came in 1940 from Mark Oliphant's laboratory in Birmingham, which was pursuing microwave development as well as uranium separation. The Australian Oliphant had learnt his direct approach to physics as a member of Rutherford's Cavendish squad, and of whom Rutherford once said, 'Oliphant will tell you what to do. He's a very fast worker.' On a US trip in 1938, Oliphant had admired the Varian brothers' klystron at Stanford and introduced this technology to the Birmingham group. Following an initial suggestion from Oliphant, John Randall and Henry Boot developed their famous cavity magnetron, producing microsecond pulses at a wavelength of 10 cm with enough power to light a cigarette. The magnetron with its cylindrical ring anode was a sort of electronic whistle, whose bunches of electrons produced powerful bursts of short-wavelength radiation. Although the powerful output of the cavity magnetron was a major breakthrough, its radiation was spread over too wide a wavelength range. Scientists found that, if the magnetron's holes were interconnected by wires, called 'straps', the electronic whistle became much more piercing. These seemingly modest developments were the technology that ensured air supremacy and prevented Britain from falling into Nazi hands. The advantages of radar could be exploited everywhere, at sea and on the ground as well as in the air, and to help focus further radar development work, what became known as the Telecommunications Research Establishment (TRE) was set up, directed by A P Rowe. After several moves, this was finally settled at Malvern to be out of range of amphibian German commando raids. The British realized that RDF, as it was still known, and the code-breaking effort to intercept and decipher enemy signals were vital and made sure both were supplied with suitable manpower. Both played a vital role in the war effort and both led to important new technology.

RDF development took place on a wide front, as systems for different applications had their own special problems. Naval radar, where systems would be away from base for months at a time, needed a different approach than those mounted in aircraft, which could be serviced every day and regularly updated. Countermeasures for enemy systems was another set of problems. A formidable array of British scientific talent was marshalled for radar work, including John Cockcroft, Norman Feather, Robert Hanbury Brown, Bernard Lovell and Philip Dee. Among those eventually assigned to radar work were Herbert Skinner from Bristol, once a student at the Cavendish Laboratory, Frank Goward, recently graduated from Cambridge, John Adams, a technical apprentice, and Mervyn Hine, another Cambridge student.

In the USA, where RDF became known as radar (radio detection and ranging), attention had also focused on the need for short wavelengths, and specialists were surprised when a British mission led by Sir Henry Tizard demonstrated the spectacular new magnetron. The USA rushed to establish a special new research centre. The result was the famous Radiation Laboratory at MIT, which by 1945 would grow to a total staff of 4000, second only to the Manhattan Project in defence-related research and development. Its director was Lee DuBridge, Chairman of the Physics Department at Rochester University and a friend of Ernest Lawrence. Supervising the actual scientific research was Isidor Rabi, a physics professor from New York's Columbia University, and early scientific recruits included K T Bainbridge, of mass spectrography fame, Luis Alvarez, who went on to win the Nobel prize in 1968 for his physics discoveries with bubble chambers, Edward Purcell, who shared the 1952 Nobel prize with Felix

Bloch, and a young New York polymath called Julian Schwinger. Because Feynman worked at Los Alamos and Schwinger at MIT during the war, their paths diverged. However, they were soon to cross in a major new scientific endeavour which ensured US pre-eminence in the immediate post-war period.

Rabi had arrived in New York City at the turn of the century as the child of an Austrian immigrant. During his statutory spell in Europe after a doctorate at Columbia, Rabi came under the influence of Otto Stern and learnt about the effect of magnetic fields on atomic beams. Returning to Columbia in 1929, Rabi went on to develop a powerful method for measuring the magnetic properties of

Figure 10: *Julian Schwinger worked at the MIT Radiation Laboratory during the Second World War and acquired a different slant on physics to his Los Alamos contemporaries.*

nuclei. In 1944, Rabi was awarded the Nobel physics prize for this work, one year after his teacher, Stern. The cauldron of New York City is a vast breeding ground for talent of all kinds, in every field of human endeavour, and Columbia acted as a magnet for city youth with intellectual ambitions. Rabi, the son of a poor immigrant who had to learn his English on the street, was one example. In the next generation of city talent came Julian Schwinger, born in 1918, who at the tender age of 17 impressed Rabi by his insight into the infamous Einstein–Podolsky–Rosen paradox, one of the numerous conundrums that all too easily arise when everyday thinking is inadvertently muddled with quantum theory. Rabi pushed Schwinger through Columbia's undergraduate school but, with the course not satisfying Schwinger's intellect, the young New Yorker amused himself by doing research and had physics papers published before he was 20. Schwinger's physics could appear like poetry, elegant and impressive, but with the underlying message not always evident. Although Schwinger and Feynman were eventually to converge on the same central physics problem, their personalities and their approaches were very different. Schwinger's first research contribution had come in 1934, when he was just 16 and had assimilated enough of the current formalism to extend the Dirac–Fock–Podolsky approach to quantum electrodynamics. At that time, Feynman was still collecting mathematical equations.

These days many people think first of Feynman's approach to modern quantum electrodynamics, because it is semipictorial and more intuitive, but Schwinger had got there first by an independent route. While Schwinger's approach was the more elegant, for beginners it was like trying to get familiar with a new city from its official history. For most people, a simplified sketch of the main streets or a subway map are a better way to start. Schwinger strived for elegance and sophistication in all that he did, his clothes, his speech, his physics. Much has been written about Feynman, who was so noticeable, a reliable source of sound bites. Schwinger could not be edited so easily and was more difficult for his contemporaries to assimilate. Schwinger strived for perfection in everything he did, whether it was a physics lecture or a research paper. Ambidextrous, he could write two different equations at the same time, an obstacle which the less gifted found difficult to overcome, especially when just one Schwinger equation could be more than a handful. Schwinger was also nocturnal, arriving in the office in the afternoon and working through the night, and therefore difficult to overlap with. Writing later about his concerted attempt to cast his theory in the most general form, Schwinger said, 'There already were visions at large, being proclaimed in a manner somewhat akin to that of the Apostles, who used Greek logic to bring the Hebrew god to the gentiles.' Contrast this logic with Feynman's laid-back approach, 'I'm always very careless, when I go on a trip, about the address or telephone number or anything of the people who invited me. I'll figure I'll be met, or somebody else will know where we're going; it'll get straightened out somehow.' If Schwinger were a concert soloist, Feynman would have been rock and roll.

At the outbreak of the Second World War, Hans Bethe at Cornell, yet to achieve US citizenship, was eager to contribute to the war effort and initially turned his attention to the physics of armour-piercing projectiles. This objective was his own choice but, even before his goals were influenced by official thinking, Bethe's mighty intellect had made important contributions. Subsequently he turned to the physics of shock waves, and then to radar, where he tried to understand how radiation was diffracted through small holes. To continue this work for the MIT Radiation Laboratory, Bethe recommended a group of five physicists, including Schwinger and the young Robert Marshak from Rochester.

When recruiting for the Manhattan Project began in 1943, it was clear that Bethe would be an early choice. Schwinger too was asked to go to the new laboratory, but preferred instead to remain at the MIT Radiation Laboratory, where he worked in the theory group under George Uhlenbeck, a Dutchman who had emigrated to the USA in 1927. During his time at MIT, Schwinger made important contributions to the theory of waveguides, the tubes and channels needed to transmit high-frequency signals. At MIT, Schwinger also profited from the university environment to keep up to date with physics and heard lectures by Wolfgang Pauli, who was sojourning at Princeton. At the end of the war, when the more isolated Los Alamos group also had to be updated, Schwinger was one of the lecturers chosen. Richard Feynman admitted to being impressed by his first encounter with his suave contemporary, who had generated a considerable reputation in physics circles.

The MIT environment clearly affected Schwinger's thinking. In his first encounters with radar problems, Schwinger later described how he initially tried to apply nuclear physics thinking. As he learned more about microwaves, he began instead to apply radiation techniques to nuclear physics. This could be one reason why the subsequent work of Schwinger, with its frequent use of Green functions and variational methods, was so difficult for the nuclear physicists to understand, and how Feynman's more intuitive approach turned out to be the more popular. Caught in a pincer movement by Feynman and Schwinger, the quantum problems which had been haunting the European theorists for almost twenty years were eventually to crumble, and the intellectual mantle that had once rested at Göttingen was inherited by sons of New City City.

At 5.30 am on 16 July 1945, night became day at the Trinity site in the New Mexico desert. A plutonium fission bomb lit up the sky brighter than any Sun. There could be no more dramatic demonstration of scientific prowess. As the sky glowed and the thunderous noise shook the Almagordo desert, the non-scientists who could not think in powers of ten were impressed by the power of science. The Los Alamos scientists had achieved their goal. The war in Europe was already over, and within weeks it would be over in the Pacific too. In the space of two years, crude bench-top experiments and blackboard diagrams had been transformed into the greatest release of energy that the planet had ever seen. The impact remained seared into the collective consciousness.

More than anything, the vast scientific research and development projects of the Second World War showed the value of close collaboration. Instead of attacking a problem piecemeal, results came by concentrating resources in or around a single major research centre, with a small number of well defined goals. If enough minds and sufficient resources were brought to bear on a central problem, it would eventually crack.

With the war over and their duty discharged, the scientific talents looked back to the pure research that they had been doing before, remembering unanswered questions and inviting problems, but their minds had been sharpened by their wartime experience. It was an experience that none of them would ever forget.

THE GRAPES OF WRATH

The physics investment of the war effort quickly brought new results in the immediate post-war period. During the Second World War, the USA had assembled at Los Alamos and other big laboratories an unprecedented concentration of scientific talent. The big names, Oppenheimer, Lawrence, Fermi, Teller and Compton, were acclaimed as national heroes who had finessed the war to its conclusion. The fruits of the wartime physics endeavour were there for all to see. However, the achievements were also the culmination of many generations of physics education which had ensured that the US universities had been well stocked with talent when the recruiting net had been cast in the early 1940s. To safeguard US interests, this physics investment had to be continued so that, if the net were to be cast again, it would pull in an adequate intellectual catch. Paradoxically, the wartime effort with its concentration on a few major strategic goals had interrupted the normal supply of fresh physics manpower. James B Conant of the National Defense Research Council estimated that the war had cost the USA many thousands of advanced science graduates, and that the former level of supply would not be re-established for almost a decade. Science education became a priority.

Before the war, government funding for university physics had been almost non-existent. Lawrence had had to rely on medical money to support much of his cyclotron work. During the war, US federal spending in scientific research had increased from US $48 million to US $500 million annually and, with the war won so convincingly, few in the USA resented such spending. Its work done, the Manhattan Project was disbanded, but in October 1945 a new body, the US Atomic Energy Commission (AEC), was established to look after the peacetime development of atomic energy. However, the Defense Department, eager to retain the nuclear initiative, outmatched the AEC in spending power. By 1949, 60% of the federal spending on physics research on US university campuses came from the Defense Department. As US physicists returned to their universities after their wartime exploits, it was soon clear that physics would receive an honourable discharge, so that resources and manpower would continue to be available on a lavish scale.

At Berkeley, even before the war, Lawrence's cyclotrons had steadily increased in size. Physics machines went from 27 inches in diameter to 37 inches and then to 60 inches, while plans had been prepared for a mighty 184-inch machine, limited only by the size of steel plates which could be milled at the time. This large machine would not fit anywhere on the Berkeley campus, and a new site had been selected up in the hills overlooking the San Francisco Bay, and 10 000 tons of concrete were poured for it in 1940. With the advent of war, obtaining the steel for this machine became a problem, but in any case there was a more fundamental limitation that it would run up against.

In 1929, it had been a paper by Rolf Wideröe, a Norwegian engineer working in Germany, which had sparked Ernest Lawrence's idea for the cyclotron. An earlier Wideröe idea had been for a 'beam transformer', in which a circulating beam of electrons acted as a secondary winding. Wideröe tried valiantly to make the idea work but was beaten by electrons piling up on the sides of the ring. Gregory Breit and Merle Tuve at the Carnegie Institution in Washington, Ernest Walton at the Cavendish Laboratory, and Leo Szilárd and J Tuck at Oxford had also tried to coax such a technique along. Such 'induction accelerators' were eventually made to work by Donald Kerst and Robert Serber at the University of Illinois in 1940, and production, under the name 'betatrons' switched to the General Electric Company. Unlike Lawrence's cyclotron, where the beam spirals out from the centre, a betatron holds its beams in a circular orbit.

In wartime Germany, Wideröe worked on betatrons, one goal being to produce powerful x-ray beams which might interfere with aircraft navigation, detonate bombs, or even kill aircrew—a 'death ray', an idea which had been considered and discarded by the British ten years before. On a more realistic level, sturdy betatron x-ray machines went on to find widespread application in industry and hospitals. In 1944, a US-built betatron was shipped to Woolwich Arsenal, London, where the original idea was to use it as a portable x-ray machine for examining unexploded bombs or for detecting flaws in armour plate. This modest machine was ultimately to become the forerunner of a new dynasty of particle accelerators.

As particle energies increased, the conventional cyclotron came up against a fundamental limitation which would have severely hampered the energy reach of Lawrence's planned 184-inch machine. As the effects of special relativity come into play, the accelerated particles start to get out of step with the applied impulses, and no more acceleration is possible. In 1943, Mark Oliphant, then working on the Manhattan Project, thought this might be overcome by suitably shaping the magnetic field and the radio-frequency of the electric field to compensate for relativity, accelerating the particles in a doughnut-shaped torus. With Oliphant distracted by wartime pressures, this 'synchrotron' idea was further developed by Edwin McMillan (who was Lawrence's brother-in-law), in Lawrence's Berkeley laboratory, and by Vladimir Veksler in Russia. Veksler and McMillan's vital innovation was the idea of 'phase stability', in which particles injected at a suitable point during the radio-frequency cycle remain bunched, giving bursts of particles.

Before such synchrotrons were built, conventional cyclotrons, using a fixed magnetic field, could overcome the relativity barrier by modulating the high-frequency electric field to keep pace with relativity, in the synchronized cyclotron or 'synchrocyclotron' method. Lawrence's 184-inch dinosaur was saved. Using a generous post-war cash injection from the Manhattan Project, the synchrocyclotron technique was applied to the giant Berkeley installation which had lain dormant during the war, and in November 1946 this machine began to supply 190 MeV deuterons, but this method had obvious limitations. The synchrocyclotron magnet has to be a flat pancake, inside which the circulating particles gradually spiral outwards. For particles to attain very high energies the magnet has to be correspondingly large and becomes prohibitively expensive. In synchrotrons, on the other hand, the magnet has only to enclose the doughnut-shaped ring in which the particles orbit, but the untried synchrotron technique had yet to be demonstrated.

In 1945, the discovery that several scientists had been passing information to the Soviets, and the 1946 MacMahon Act effectively stopped the free exchange of nuclear information between US and UK scientists. As British scientists returned home after their wartime work in North America, Anglo-American collaboration in physics looked to be seriously compromised. Undeterred, the UK physicists set out to develop their own programme. The plan was to set up a new atomic energy research laboratory at Harwell, a former wartime airfield near Oxford, under the directorship of John Cockcroft. Harwell would be the site for new research reactors and particle accelerators. To prepare the way, a panel on accelerating particles was set up under Oliphant, with representatives from universities and industrial firms, and a group at the TRE, Malvern, began to plan for new Harwell equipment. In charge of this physics was Donald Fry, and a member of his team was Frank Goward, who had graduated from Cambridge directly into wartime radar work, moving to Malvern in 1942 to work on ground-based antennae.

The energetic Goward realized how the electron betatron at Woolwich could be adapted into a McMillan–Veksler synchrotron, and with D E Barnes of Woolwich in August 1946 succeeded in doubling the machine's nominal betatron electron energy of 4 million electronvolts (i.e. 4 MeV). The modest machine was subsequently moved to Malvern, where it eventually achieved 14 MeV. In October of that year, an electron synchrotron built by the US General Electric Company, who had previously been building betatrons under Kerst's direction, reached 70 MeV. The first synchrotrons had arrived.

Cockcroft's original plan for the new Harwell laboratory was based on traditional pre-war nuclear physics lines, a sort of super-Cavendish Laboratory. With his nuclear physics background, Cockcroft placed high priority on a machine for producing high-energy proton beams but, as he still had to wind up

Figure 11: *The first synchrotron. Following the invention of the synchrotron idea by Vladimir Veksler in the USSR and Edwin McMillan in the USA, in 1946 Frank Goward and D E Barnes at the TRE converted this bench-top betatron accelerator, imported from the USA, into an electron synchrotron, doubling the machine's output energy.*

business at the Chalk River laboratory in Canada, he appointed Herbert Skinner from TRE to head Harwell's Physics Division. Among those transferring to the new accelerator business were Donald Fry, Frank Goward and John Lawson from TRE, together with Gerry Pickavance, who had taken over responsibility for the Liverpool cyclotron when Chadwick departed on war business. To supervise the engineering of the new machine, Skinner invited the young John Adams from TRE. The 110-inch Harwell synchrocyclotron, capable of 175 MeV, then the highest-energy machine in Europe, came into operation in December 1949.

However, in the USA, flush with post-war funds, new particle accelerators were springing up like mushrooms. At Berkeley, McMillan had ambitious plans for an electron synchrotron, while Luis Alvarez pushed ahead with linear accelerators for protons, a giant version of Wideröe's 1927 idea. At Cornell, Robert Wilson, returning from Los Alamos, built an electron synchrotron.

Cyclotrons at Princeton and the University of California at Los Angeles (UCLA) were converted to synchrocyclotrons. Other electron synchrotrons were either operational or soon to be so at General Electric, Ames (Iowa), MIT and Purdue. As well as at Berkeley, US synchrocyclotrons were built at Rochester, Harvard, Columbia, Chicago and the Carnegie Institute. In Canada, one was built at McGill, Rutherford's former home in Montreal. (Cyclotrons, many dating from the pre-war era, were too numerous to mention. A 1948 survey listed 21 machines in the USA and 14 in Europe.)

New laboratories, too, appeared. Los Alamos and Oak Ridge came into existence during the war years. Chicago's Met Lab was established in a new home at Argonne. Berkeley had undergone a major transformation on its new hilltop home. At Columbia, Isidor Rabi, returning from the MIT Radiation Laboratory, and Norman Ramsey, returning from Los Alamos, found themselves suddenly cut off from ongoing research. In their view, Columbia had played an important role in getting the US fission effort off the ground; Columbia had been the initial US base of Fermi and Szilárd. There were new research reactors at Oak Ridge and Argonne, but neither of these was convenient for New York. Berkeley had been elevated to superstar status. Cambridge, Massachusetts, home of both MIT and Harvard, was also pushing for its own accelerator laboratory, but the initiatives soon decided to join forces under a single banner. Representatives from Columbia, Cornell, Harvard, Johns Hopkins, MIT, Pennsylvania, Princeton, Rochester and Yale formed 'Associated Universities, Inc' and, with funding from the interim Manhattan Project–AEC, established a new laboratory, Brookhaven, at Camp Upton. This former army base on Long Island had been set up in the First World War as a staging post for US troops *en route* to the battlefields of Europe, a role which was re-enacted during the Second World War. An initial plan to build a 700 MeV cyclotron at Brookhaven was dropped when a similar project, the Nevis cyclotron, was pushed through at Columbia.

The synchrocyclotron technique, calling for a large circular magnet, could not be continued indefinitely, and for higher energies the synchrotron was the only route. In 1948, the AEC recommended that two large proton synchrotrons be built: one at Brookhaven and the other at Berkeley. In the UK, Oliphant as usual had been extremely active and had launched a project to build a major proton synchrotron at Birmingham. Before this project was complete, he returned to his native Australia, where he immediately launched a similar project for the Australian National University, Canberra, which was never completed.

In California, the big new linac built by Luis Alvarez at Berkeley prompted Bill Hansen at nearby Stanford to see whether something similar could be done for electrons. During the war, Stanford's microwave expertise had been exported to the East Coast of the USA to help to launch the Sperry Gyroscope Company into electronics for radar systems. During the war, the power output from klystrons improved dramatically. While continuing to develop high-power klystrons, the post-war Stanford laboratory had its first success in 1947 accelerating a linear beam of electrons to 6 MeV in 12 ft (3.7 m) using a 1 MW magnetron. This was the Stanford Mark I machine, and Bill Hansen set his sights on a 160 ft (50 m) billion volt machine. *En route* to such a big machine, the 33 MeV Mark II electron linac was operated in 1949 with a single huge klystron. The first 30 ft (9 m) module of the Mark III machine was powered to 75 MeV in 1950. By December 1957 it had grown to more than ten times its original length and routinely delivered 900 MeV. With all its klystrons working, it could be coaxed to the magic gigaelectronvolt threshold. Stanford was firmly set on a linear electron path.

While all these new techniques were being refined and impressive machines were being built, the next physics discoveries came instead from natural cosmic rays. In 1932, Carl Andersen at Caltech had discovered Dirac's positron, the antimatter counterpart of the electron. Continuing with his cosmic-ray research with Seth Neddermeyer, in 1936 he discovered another new particle, heavier than the electron but lighter than the proton. Anderson called the new particle the 'mesoton' from the Greek 'meso', meaning 'between', but the name was subsequently altered to 'mesotron' and finally abbreviated to 'meson'. At the same time, Hideki Yukawa in Japan predicted that the nuclear force which bound protons and neutrons in nuclei and which overcame the enormous electrical repulsion between the closely packed protons was carried by an as yet unseen particle. The quest for the Yukawa particle and a push to develop a satisfactory theoretical framework for it was to dominate physics for the next thirty years.

Anderson's meson was immediately suspected to be the Yukawa force carrier, as it had the correct mass. If it were Yukawa's nuclear force carrier, it had to interact with nuclear material. An experiment in Rome in 1943 by three Italian physicists, Marcello Conversi, Ettore Pancini and Oreste Piccioni, carried out against a backdrop of wartime bombing, showed that, although the meson decayed on a time scale compatible with being a Yukawa particle and could be stopped in iron, the meson passed relatively easily through other materials. If the meson were a nuclear particle, it should interact with all kinds of nuclei. News of this experiment reached Enrico Fermi in the USA, and a historic analysis by Fermi, Edward Teller and Victor Weisskopf showed that there was a million-millionfold discrepancy between the nuclear affinity of the cosmic ray meson and that expected from a Yukawa particle.

The classic tool for discovering new particles was the cloud chamber, invented by C T R Wilson some fifty years previously at the Cavendish Laboratory, but some work had been done using blocks of photographic emulsion in which traversing particles would leave tracks. Blocks of emulsion were also more portable and manageable than cloud chambers. At Bristol, a group led by Cecil Powell, who had once worked at the Cavendish Laboratory, began to use photographic emulsions to record the tracks of cosmic-ray interactions at high altitudes. In a set of cosmic-ray exposures made 2800 m up in the French Alps, Powell's team identified more than 50 meson tracks, two of which were curiously kinked, with one meson decaying into another. Soon more such tracks were seen, and the observations were published in the journal *Nature* in May 1947.

Early in 1947, J Robert Oppenheimer, who had been appointed Director of Princeton's Institute for Advanced Study, moved for a topical meeting to focus attention on new theoretical physics issues. This was held at Shelter Island, off New York's Long Island, at the beginning of June. One of the controversies discussed at the meeting was the Italian meson findings. Because journals were seldom sent by airmail in those days, news of the Bristol group's latest discovery had not reached the USA in time for the meeting. However, Robert Marshak and Hans Bethe speculated that perhaps there were two cosmic-ray mesons. One was the Yukawa particle, with a strong nuclear affinity and produced in primary cosmic-ray collisions high up in the atmosphere. This, they suggested, subsequently decayed into a second, much less reactive particle, seen at lower altitudes. Marshak and Bethe proposed that it was this second meson which had been discovered first by Anderson. On learning of the Bristol sighting of one meson decaying into another, they quickly published their explanation. The Bristol emulsion features were variously described by the Greek letters pi, rho and sigma but, soon realizing that they were all the same, Powell preferred the pi label. Hence the name pi-meson, or pion, for the nuclear particle predicted by Yukawa. The second meson, lighter and less reactive, into which the pion decayed became known as the mu-meson, or muon. (This idea of two sequential cosmic-ray mesons had been first thought of in Japan, but the wartime communications clampdown prevented the news from reaching the West.)

Shelter Island had focused attention on other interesting new results. The US radar-oriented microwave work centred at the MIT Radiation Laboratory had been wound up after the war, and the physicists who had been drawn into this major effort dispersed to their peacetime laboratories. Immediately they were able to apply their new knowledge and techniques to problems which had long been on their minds, and fresh results were not long in coming. After Otto Stern and Isidor Rabi had pointed the way to make measurements of nuclear magnetism, in 1946 Felix Bloch at Stanford and Edward Purcell of Harvard increased the precision of these measurements to obtain values for the magnetic moments of nuclear particles. Columbia, with Rabi, was a centre of excellence for microwave work. Even during the war it had operated as a satellite of the MIT laboratory. The same year, Willis Lamb at Columbia, working with Robert Retherford, who had worked on vacuum tubes during the war, set out to use

Figure 12: *US physicist Isidor Rabi (right) received the Nobel prize for physics in 1944 for his work on atomic and molecular beams. Later he pushed for the creation of the Brookhaven National Laboratory as a regional US research centre. Extending this idea to the international arena, at the UNESCO General Conference, held in Florence in June 1950, he tabled the idea which went on to become the CERN laboratory. With Rabi is Cecil Powell, who led the group at Bristol which in 1947 discovered the pion in cosmic rays. Powell went on to receive the 1950 Nobel prize.*

microwave techniques to measure the fine structure of the hydrogen spectrum. On 26 April 1947 their experiment revealed that certain pairs of hydrogen spectral lines, which according to ordinary quantum mechanics were expected to have the same energy, were in fact slightly displaced. More such precision measurements were made at Columbia by John Nafe and Edward Nelson. Polycarp Kusch and Henry Foley, also at Columbia, used these refined microwave techniques to measure the magnetic moment of the electron.

A rotating electric charge behaves like a magnet, and the spinning electron was no exception. However, the experiments by Otto Stern and W Gerlach in the 1920s had shown that the spinning electron has twice the magnetism expected by a classically spinning charge. This mysterious factor of two had been impressively explained by Dirac's triumphal relativistic equation for the electron, but Kusch's experiment showed that the electron's magnetism is in fact 0.1% different from the Dirac value. For these discoveries, which showed that the quantum behaviour of the electron needed further explanation, Bloch and Purcell earned the 1952 Nobel prize, while Lamb and Kusch made the trip to Stockholm three years later.

At Shelter Island, the news of Lamb's tiny discrepancy (the 'Lamb shift') had been leaked beforehand. In any case a pre-war experiment by Simon Pasternack had suggested such an effect. Suspecting that it was due to electrodynamics, Hans Bethe did an immediate back-of-the-envelope calculation which appeared to be on the right track but did not fully take into account relativity. Norman Kroll and Willis Lamb were the first to produce a treatment which included relativity, but it was a one-off calculation and not a general scheme for doing things. Julian Schwinger, also at Shelter Island, realized that explaining the new effect called for a major overhaul of quantum electrodynamics.

At Shelter Island, Isidor Rabi explained that experiments at Columbia had detected magnetic effects in the spectroscopy of hydrogen and deuterium which were slightly too large for the accepted theory. Suspicion soon centred on the electron's magnetism as the culprit but, rather than confronting the new problem immediately, Schwinger preferred to go on his honeymoon that summer (as he put it, 'I abandoned my bachelor quarters and embarked on an accompanied, nostalgic trip'). However, by the end of the year, Schwinger submitted a paper to *Physical Review* which explained exactly how the magnetism of the electron is altered by the electron's subtler interactions—when the electron 'talks to itself'. It contained an important new development.

Since the time of Dirac, physicists trying to formulate a theory of quantum electrodynamics had been dogged by troublesome infinities which made their calculations seize up. There seemed to be no way of avoiding these obstacles, and the mood began to shift towards despair. When they recommenced their work after the Second World War, theoretical physicists knew that the infinities would still be waiting there to haunt them, but Schwinger saw how they could be avoided if they were handled carefully. Quantum electrodynamics was 'renormalized' (see page 107).

While Schwinger was formulating his new theory, an English mathematician called Freeman Dyson came to Cornell as a student. Recognizing talent, Bethe immediately assigned him to work out the spectroscopic discrepancy reported by Willis Lamb, but at Cornell Richard Feynman had also locked onto this problem. When Feynman's mind seized a problem, the problem gradually succumbed under the constant pressure, like a boa constrictor crushing its prey. Like Schwinger, Feynman did not want to do an isolated calculation to explain a new result. Instead he wanted to recast the whole of quantum electrodynamics, which had been creaky ever since the time of Dirac, taking it apart bit by bit, and putting it back together and lubricating it to ensure it worked smoothly.

On 31 January 1948 the American Physical Society's meeting in New York was packed to the doors with scientists expecting to hear a virtuoso performance from Julian Schwinger. Enrico Fermi came all the way from Chicago and took notes during Schwinger's talk. Few had seen Fermi take such copious notes before. After the meeting, Fermi went back to Chicago with his pile of Schwinger lecture notes and began to study them, distributing them also to his Chicago colleagues. Gradually they began to understand what Schwinger was doing. On that January morning, not everybody who had wanted to hear Schwinger had been able to get into the lecture hall, and a repeat talk was arranged at nearby Columbia University. Schwinger's new approach was the talk of physics, but as yet few could understand it.

As the Shelter Island meeting had appeared to be so productive, a second meeting was organized in March 1948, this time in the Pocono mountains, Pennsylvania. Paul Dirac and Niels Bohr came from Europe. Schwinger's presentation, claiming 'relativistic invariance' and 'gauge invariance', two criteria subsequently present in every major new physics theory, was widely admired. Feynman, speaking afterwards, went down less well. Dirac and Bohr did not seem to understand what he was saying. Afterwards Feynman conceded, 'I had too much stuff. My machines came from too far away.'

That summer, Dyson and Feynman drove across country together. Dyson wanted to see some of America before going on to Princeton's Institute of Advanced Study that fall but on the trip profited from a hefty exposure to Feynman. There was the gospel according to Julian Schwinger, and the gospel according to Richard Feynman. Before arriving at Princeton, Dyson submitted to *Physical Review* his own gospel of quantum electrodynamics, which filled in the physics which Schwinger had not shown, and the mathematics which Feynman had not bothered to display. This Dyson treatment helped to establish renormalized quantum electrodynamics as a standard theory. There had been another version of the gospel, developed independently by Sin-Itiro Tomonaga in Japan. Tomonaga, like Hideki Yukawa, graduated from Kyoto University in 1929, and from 1937 to 1939 worked with Heisenberg at Leipzig. Tomonaga, like many of his US and British physics contemporaries, was diverted into wartime microwave research. During this time, a German U-boat delivered to Japan copies of a document classified as 'top secret ', a new paper by Werner Heisenberg on the scattering matrix (S matrix). Unlike Schwinger and Feynman, Tomonaga was not able to learn immediately about the new microwave measurements of hydrogen spectroscopy but nevertheless was able to formulate his own version of renormalized quantum electrodynamics which paralleled that of Schwinger. Both Tomonaga and Yukawa visited Princeton's Institute for Advanced Study as Oppenheimer's guest in 1949.

Although they fell on unfertile ground at the Pocono meeting, Feynman's ideas gradually took root. The history of electrons and positrons could be written as compelling little pictures—'Feynman diagrams'—which were to become the subway maps of subatomic physics. The electrons and positrons are lines but are endowed with arrows, showing in which direction time runs. Photons, the quanta of radiation, look the same whether they are running forwards or backwards in time and are shown as wavy lines. The three dimensions of space and the fourth dimension of time cannot be accommodated on a two-dimensional sheet of paper. Feynman diagrams represent a simplified world of one space dimension and one time dimension. (These ideas had been introduced several years before by the Swiss physicist Ernest Stückelberg, but he had not introduced all the necessary ideas of renormalization.) However, the space and time dimensions

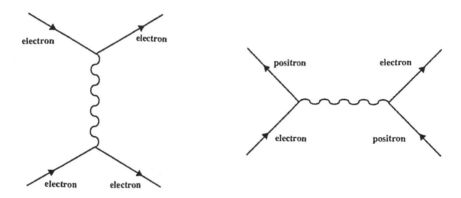

Figure 13: *Feynman diagrams use a two-dimensional space in which the horizontal axis represents the passage of time with the vertical axis as a single space dimension. The left-hand diagram shows two electrons interacting via the exchange of a photon (wavy line). The arrows on all the electrons point forwards in time. Turning the diagram through 90° (right-hand diagram), the arrows on some electrons now point backwards in time; these are now electron antiparticles, positrons. The diagram represents an electron and a positron annihilating into a puff of radiation, later to rematerialize as another electron–positron pair. Julian Schwinger said of these diagrams 'like the silicon chip, they brought calculations to the masses'.*

can be interchanged; on this relativistic map, north-south travel is just as easy as west-east. By turning the map, physics processes that pre-1947 were viewed as totally different now became intimately related. Thus the scattering of an electron is the cousin reaction of an electron and a positron annihilating into a puff of radiation, later to rematerialize as another electron-positron pair. Time travel had arrived.

In 1965, the Nobel prize for physics was awarded to Richard Feynman, Julian

Schwinger and Sin-Itiro Tomonaga for their development of quantum electrodynamics. According to the rules, the Nobel prize cannot be shared by more than three people. The development of the new quantum electrodynamics in 1947–8 meant that Europe, where the Ancient Greeks had first speculated about the ultimate division of matter, and where Planck, Einstein, Bohr, de Broglie, Schrödinger, Heisenberg, Pauli and Dirac had pieced together the quantum picture, had lost its place as the traditional home of physics theory.

On the experimental front, and despite the assembly of American particle accelerator heavy artillery, Europe still had important experimental contributions up its sleeve. In 1947, George Rochester and Clifford Butler from Blackett's laboratory in Manchester stumbled across some cosmic-ray photographs showing unexpected forked tracks. Blackett had become Professor at Manchester in 1937, after Bragg had moved to the Cavendish Laboratory following the death of Rutherford, and had established a dynamic cosmic ray group, which also included Bruno Rossi, Lajos Janossy and Bernard Lovell.

Figure 14: *Patrick Blackett, who rose to fame at Rutherford's laboratory in Cambridge, won the Nobel Prize for physics in 1948. In 1947, cosmic-ray cloud chamber tracks recorded at Blackett's laboratory in Manchester saw the first examples of 'strange' new particles which were to dominate physics thinking for many years.*

Lovell was directed into wartime radar work and went on to become director of the big Jodrell Bank radio telescope near Manchester. Resuming research after the war, Rochester and Butler redesigned Blackett's large cloud chamber and were rewarded with two pictures of unexplained heavy 'V particles' as Blackett called them. One, discovered in 1946, was electrically neutral, later identified as what is now called a K meson, or kaon; the other, discovered the following year, was a charged particle, later identified as an electrically charged version of the K meson. However, further exposures of the Manchester cloud chamber failed to

Figure 15: *French cosmic-ray physics pioneer Louis Leprince-Ringuet. Students from his influential school went on to become important personalities.*

reveal more V particles and the group began to have doubts. Rochester attended a cosmic ray meeting at Caltech in 1948 to mark Robert Millikan's eightieth birthday but adopted a low profile as no more V particles had been seen. Louis Leprince-Ringuet from France was more vociferous about a heavy meson discovered in cosmic rays and which he called the tau.

Blackett realized that cosmic ray work could benefit from exposures at higher altitudes to combat absorption in the atmosphere, and a project was launched for a new cloud chamber to be installed 3500 m up at the Jungfraujoch observatory in the Swiss Alps. This project took longer than expected; so Occhialini suggested to Blackett that the existing Manchester cloud chamber be moved to a site at the Pic-du-Midi in the French Pyrenees at a height of 2867 m. This new installation ran for emulsion experiments during the winter of 1949–50 but the cloud chamber did not arrive until the spring of 1950. In November 1949 a letter had arrived at Manchester from Carl Anderson who had seen 30 new examples of V particles in experiments at sea level and 3200 m up on White Mountain, California. The Manchester chamber at the Pic-du-Midi was followed by a larger one from the Paris Ecole Polytechnique.

These new cosmic-ray discoveries were complemented by sea-level observations by an Indiana group using a cloud chamber. From 21 000 photographs, the group found 38 examples of V particles, showing that they produced both hyperons (heavy nucleons) and a lighter particle, the theta, decaying into a pair of pions, and now known to be a neutral kaon.

The Jungfraujoch was no stranger to cosmic-ray experiments; Louis Leprince-Ringuet had carried out measurements there before the war. Subsequent installations in the French Alps continued to take photographs during the war years, and a 1946 Leprince-Ringuet photograph from the Aiguille du Midi station near Chamonix at an altitude of 3600 m first suggested the existence of a heavy meson some 1000 times heavier than the electron. Run by the Ecole Polytechnique, the station continued to take photographs until 1955, providing valuable information on K mesons and on heavier V particles, and training several generations of French researchers.

Although the flux of cosmic rays increases with increasing altitude, many of the major cosmic-ray discoveries were nevertheless made at sea level. The arduous conditions of the high-altitude laboratories could compromise the higher particle flux. One of the early post-war exponents of cosmic-ray studies, Don Perkins, who saw initial evidence for pion tracks while working at London in 1947, later pointed out that cosmic-ray research using photographic emulsions in the late 1940s and early 1950s led to the first large-scale international physics collaborations. For example a 1953 project based on Sardinia involved 22 laboratories from 12 countries. In this way small groups with meagre resources were, and still are, able to contribute to forefront research. Cosmic rays are equally accessible anywhere on the planet, and the experimental resources needed are relatively modest. By the early 1950s, cosmic-ray groups had been set up all over the world.

These post-war cosmic-ray discoveries were emphasized by the award of the Nobel prize to Blackett in 1948 and to Powell in 1950. This harvest of V particles, together with the appearance of the pion, were to be the last major particle physics discoveries made in Europe for the next quarter of a century, but this cosmic-ray windfall had been fortunate for Europe, because in 1946 the Berkeley 184 inch synchrocyclotron began to produce 190 MeV deuterons and 380 MeV alpha particles. Although strictly speaking these energies are insufficient to liberate pions from protons at rest, nuclear protons have additional energy, the 'Fermi energy', as they jiggle around inside their nuclear habitat. With this energy bonus, pions could be created. Initially they were not seen because the Berkeley researchers were preoccupied with getting their synchrocyclotron to work. Initial attempts to see pions at the Berkeley synchrocyclotron were unsuccessful, but progress began to be made when Cesare Lattes, a member of Powell's Bristol group, installed an emulsion target in the synchrocyclotron. These results were announced at the 1948 Pocono meeting, but some subsequent reports were garbled because of the parallel question of differentiating the new pion from the cosmic-ray muon.

The pion seen in the Bristol emulsions was an electrically charged particle. However, in the late 1930s, in the wake of Yukawa's prediction that a nuclear force-carrying particle had to exist, Nicholas Kemmer at Edinburgh had looked at the electric charge possibilities for such particles. Because the proton and the neutron carry different electric charges, so the Yukawa particle has to carry electric charge. However, in an elegant argument, Kemmer showed that, if the nuclear forces due to the Yukawa particle were to be charge symmetric, then the Yukawa particle also had to exist in an electrically neutral form. Such a particle should decay into a pair of photons, making it difficult to detect in traditional

Figure 16: *The Cosmotron at the Brookhaven National Laboratory, New York, was the first man-made machine to supply particles with energies comparable with those of cosmic rays. It achieved its design energy of 3.3 GeV in January 1953.*

cosmic-ray studies. In April 1950, Jack Steinberger, Wolfgang Panofsky and Jack Steller, working at McMillan's 300 MeV electron synchrotron at Berkeley, saw effects due to a neutral pion decaying into two photons. Some twenty years after Lawrence invented the cyclotron, this was the first particle to be discovered in an accelerator.

As well as discovering new particles, the new machines made the study of subnuclear particles a much more quantitative science. By 1952, the synchrocyclotrons at Berkeley, Rochester and Columbia had clarified the properties of pions. As frequently happens in physics, these particles soon ceased to be valuable experimental prizes, becoming instead experimental tools, first at the Berkeley synchrocyclotron and then at Brookhaven, where the world's first proton synchrotron began to operate in 1952. The first man-made machine to supply particles with energies comparable with those of cosmic rays, this machine was christened the Cosmotron, and achieved its design energy of 3.3 GeV in January 1953. This machine provided physicists with a new diet, scattering high-energy pion beams from nuclear targets, and Cosmotron experiments showed that the nuclear proton and neutron had cousin states at higher energy. In April 1953 the first man-made V particles were reported at the Cosmotron. In his conclusion at the conference in July 1953 held at Bagnères, near the Pic-du-Midi laboratory, Cecil Powell said, 'We have been invaded ... the accelerators are here.'

In 1950, Robert Marshak at Rochester organized the first of a series of conferences which to this day continue to provide a regular stage to discuss and examine progress in particle physics. In the late 1940s, the Shelter Island and Pocono meetings had concentrated on theoretical developments. Marshak proposed instead that the Rochester meeting should give equal emphasis to cosmic-ray work, accelerator studies and theory. In the space of a few years, the cosmic ray component withered, as the future clearly lay with the big new machines. In the early 1950s, the formidable array of synchrocyclotrons in the USA was complemented elsewhere in the world by those at Harwell, by the 156 inch synchrocyclotron at Liverpool which began operations in 1954 and could attain 400 MeV, by a small machine in Amsterdam designed by Cornelis Bakker at Philips, a 200 MeV machine in Uppsala, Sweden, and a 650 MeV machine in Russia. There were British electron synchrotrons at Glasgow (300 MeV) and Oxford (140 MeV). In Russia, Veksler, one of the architects of the synchrotron principle, built the first of a series of electron synchrotrons. On the proton synchrotron front, the Brookhaven Cosmotron was soon joined by the 1 GeV machine at Birmingham, UK. Launched by Oliphant as a 1.3 GeV design, this subsequently had been compromised by design and construction problems. Its restricted environment did not permit the range of experiments which such a machine in principle could support. The beam pointed towards the ladies' toilet and concrete blocks had to be brought in to prevent the ladies from being irradiated. Bigger and better-designed machines were under construction at Berkeley, and at Dubna, near Moscow.

The race to higher energies had begun, but the leading runners were almost all American. In Europe, there was still a considerable effort in the UK, but Britain was already out of breath before the race had begun in earnest. For a while it looked as though the Americans would have the field to themselves.

ALL WE WANT IS THE WORLD'S BIGGEST MACHINE!

Science, by its very objectivity, acts as a common bond between nations. From time to time, scientific projects become sovereignty issues: discovering new chemical elements in the nineteenth century and naming them as national landmarks, or putting men on the Moon in the twentieth century. Having done so, the chemical elements or the Moon cannot be patented. In every country, scientists look at the same problems in the same way, eager to acquire the same knowledge. If one country hushes up its research findings, as in the Manhattan Project, the same knowledge will ultimately be acquired elsewhere. Scientific curiosity cannot be quenched.

But the wartime efforts had accelerated a trend in science. The era of ingenious tabletop experiments, Rutherford style, was almost over. Taming fission and microwaves had demonstrated that unprecedented resources could be needed to crack new scientific problems. Wartime experience had also shown the value of large-scale collaboration. The whiff of post-war oxygen stimulated several new directions. The rapid emergence of nuclear power as a strategic and economic factor quickly led to the establishment of atomic energy commissions in the major countries. To do their work, these commissions needed access to research laboratories: in the USA at Los Alamos, Brookhaven, Berkeley, Oak Ridge, etc; in the UK at Harwell; in France at Saclay; in Russia at Moscow and Dubna. The United Nations had come into being and was setting up specialist executive organizations, including what was to become the International Atomic Energy Authority, and the United Nations Educational, Scientific and Cultural Organization (UNESCO).

In Europe, traditional militant national pride, tempered by two world wars, had given way to a new awareness of belonging to a continent. Embarrassed by having caused so much strife and inflicting it on the rest of the planet, Europe felt that it had to present a more united front to the rest of the world. On 19 September 1946, Winston Churchill called for the establishment of 'a sort of united states of Europe'. From the Organization for European Economic Cooperation, established in 1948, grew the Council of Europe, and the pragmatic European Coal and Steel Community, the forerunner of today's European Community.

In the wider context of the UN, an early French proposal had suggested setting up UN research laboratories. This would foster the establishment of an international spirit, with science providing a framework in which international collaboration could produce tangible results, rather than just providing a forum for discussion and negotiation.

The enforced emigration of Jewish scientists from mainland Europe in the 1930s had been to the lasting benefit of several countries, but the USA benefited the most, amplifying this timely injection of talent by its own mighty resources. With the USA clearly identified as the leading post-war scientific power and providing a natural intellectual focus, scientific emigration from Europe continued in what would eventually be called the 'brain drain'. This trend had to be stemmed if the Old Continent was not to find itself starved of new talent. This had motivated the Swiss writer Denis de Rougement, soon to found the European Cultural Centre, who discussed with Einstein the effectiveness of linking the emerging pan-Europeanism with plans to develop the peaceful applications of nuclear energy.

One outcome was the European Cultural Conference in Lausanne in December 1949, which set the stage for new developments. This meeting, attended by 170 influential people from 22 countries, was one of four specialized meetings following the Congress of Europe in the Hague in May 1949. In his report at Lausanne, Denis de Rougemont deplored an increasing trend towards secrecy in nuclear physics and advocated the development of a European outlook. Specifically he moved for the creation of a 'European centre for atomic research'. Such ideas had been touched on earlier, notably the formal inauguration in November 1949 of the new Amsterdam synchrocyclotron, but the international audience at Lausanne was to prove more receptive. To set the tone for the meeting, Raoul Dautry, Administrator-General of the French Atomic Energy Commission and President of the French Executive Committee of the European Movement, read a message from the French physicist Louis de Broglie. Unable to attend personally, de Broglie, winner of the 1929 Nobel prize for his elucidation of the wave nature of particles, maintained that scientific collaboration between a large number of European countries could undertake projects which, by their cost and implied resources, were beyond the means of individual nations. The objectivity of science, in contrast with other fields, should facilitate such an endeavour, de Broglie concluded. Important messages attract additional significance when signed by Nobel prize winners, and this was no exception. Following up the keynote speech with his own ideas, Dautry affirmed that astronomy and astrophysics on the one hand, and atomic energy on the other hand, would be ideal vehicles for such international collaboration.

To help him to achieve these objectives, Dautry could call on powerful colleagues in France: Francis Perrin, soon to become High Commissioner of the French Atomic Energy Commission, and the distinguished figure of Pierre Auger, who, like Perrin, had made important contributions to atomic and nuclear

Figure 17: *The distinguished French physicist Pierre Auger made important contributions to atomic and nuclear physics in the 1930s. Auger, full of enthusiasm and ideas in his new role as UNESCO's Director of Exact and Natural Sciences, was to play a key role in the events leading to the establishment of CERN.*

physics in the 1930s. Auger, full of enthusiasm and ideas in his new role as UNESCO's Director of Exact and Natural Sciences, was to play a key role in the ensuing developments, with UNESCO providing a natural stage. Another prominent French player was Lew Kowarski, who had helped to pioneer nuclear fission and had become a director at the French Atomic Energy Commission. Kowarski, who had worked on British fission projects during the war, understood the special position of the UK. Although Britain had a head start, the country had nevertheless not been able to capitalize on its close wartime links with the USA and was trying to 'go it alone'. Other European countries, notably the Netherlands and the Scandinavian nations, had strong links with the UK and could be drawn into a UK-dominated effort. At the same time as trying to go it alone, the British knew that they were part of Europe. International ventures put forward in Europe had at least to invite British participation at an early stage. It would be up to the British to decide whether they joined or not.

At the UNESCO General Conference, held in Florence in June 1950, the seed which had been planted at Lausanne still lay dormant. Among the US delegation at Florence was 1944 Nobel physics prize winner Isidor Rabi. Born on 29 July 1898 in Rymanov, Galicia, and emigrating with his parents as a small child to the USA, Rabi had been educated as a chemist at Cornell, and after professional work as a chemist subsequently turned to physics at Columbia University, New York. Apart from a spell at the MIT Radiation Laboratory during the Second World War, Columbia was Rabi's academic home for the rest of his life. Physically small and invariably smartly dressed, Rabi related that, on a cab ride home from New York's La Guardia airport, an inquisitive New York cab driver, proud of his deductive powers, immediately asked whether his fare was a college professor or a lawyer. On learning that the former was right, the cab driver replied that college professor had been his initial impression, but he had subsequently changed his mind as Rabi was 'too well dressed'.

In his wartime post at the MIT Radiation Laboratory, Rabi had been impressed by the fruitful and open collaboration with the British, who had disclosed all their new microwave secrets to the USA, and had also brought their experience of operational warfare. In his book *The Scientist and the Statesman*, Rabi related that he was also impressed by the sheer vitality and energy of the British approach to wartime technological challenges. Overcoming obstacles requires energy. Rabi says, 'We made many mistakes but, because of the great energy which was applied to the task, these mistakes were made quickly and soon corrected. When one moves at high speed a detour does not lose as much time as an orderly approach which is over-timid and over-cautious.' Rabi also affirmed that any joint enterprise involving scientists has to be one of mutual respect and deplored the widespread attitude that 'the scientists should be on tap but never on top'. 'Wartime experience,' said Rabi, 'moulded the post-war USA and pointed to new requirements in a way which did not happen so quickly in Europe.' Rabi had quickly pushed, with Norman Ramsey, for the establishment of a major new US research laboratory, Brookhaven, on New York's Long Island, on the East Coast. Brookhaven, in Rabi's mind, was a role model for what could be achieved elsewhere. The Long Island laboratory was a collaboration between major university partners. Substituting nations for universities gave an immediate template for an international laboratory.

On arriving at Florence, Rabi was surprised to discover that the agenda for the UNESCO meeting included no mention of the European nuclear physics collaboration which had been mooted at Lausanne, despite the fact that UNESCO had UN responsibility for these matters. Setting up physics laboratories was something Rabi knew about, but international committee work was not. The first thing to do was to get an item introduced on the agenda, despite the apparent indifference of his colleagues on the American delegation, who could not appreciate what all the fuss was about. More helpful to Rabi were Pierre Auger and the Italian physicist Edoardo Amaldi, who had worked in Fermi's Rome laboratory before the war. Amaldi, who had been offered a post

Figure 18: *The other founding father of CERN was Edoardo Amaldi, who had worked with Enrico Fermi in Rome in the 1930s. Although Amaldi had been offered a post in the USA by Fermi, he nobly preferred to stay in his native country and to strive to restore Italian physics after the chaos of the war, an objective in which he succeeded admirably. Amaldi went on to become one of Europe's great post-war scientific statesmen, his achievements appearing to stem from a deep sense of duty rather than personal ambition.*

in the USA by Fermi, had nobly preferred to stay in his native country and to strive to restore Italian physics after the chaos of the war, an objective in which he succeeded admirably. Amaldi also went on to become one of Europe's great post-war scientific statesmen, his achievements appearing to stem from a deep sense of duty rather than personal ambition. He was one of the few who could clearly see, and achieve, a perfect balance between national and international interests.

In the late 1940s, Amaldi had cultivated good contacts in Britain, and, sensitive to the possibilities of international collaboration, had become Vice President of the International Union of Pure and Applied Physics (IUPAP). Drafted with the assistance of Auger and Amaldi, the proposal from Rabi at Florence requested UNESCO 'to assist and encourage the formation and organization of regional research centres and laboratories in order to increase and make more fruitful the international collaboration of scientists in the search for new knowledge in fields where the effort of any one country in the region is insufficient for the task'. Presenting the motion, Rabi pointed out that the initiative 'was primarily intended to help countries which had previously made great contributions to science' and that 'the creation of a centre in Europe ... might give the impetus to the creation of similar centres in other parts of the world'. Aware that physics could quickly become the monopoly of a small club of world powers, Rabi advocated that the new centre should focus primarily on physics research. The motion was unanimously accepted. Where Europeans had failed to reach a consensus, an American resolution for Europe at a meeting of a UN agency had opened a new door. Europeans now had to explore what lay behind it.

The original Rabi proposal was sufficiently flexible to have led in many directions, but Rabi had given additional hints after the meeting, even suggesting that US contributions could be forthcoming. This is not surprising at the time; the Marshall Plan to rebuild Western Europe was to contribute US $13.5 billion between 1948 and 1951. In such a context, a new physics centre would be a drop in the ocean. The two men who took Rabi's baton and sprinted with it were just those who had helped him draft the Florence resolution— Edoardo Amaldi and Pierre Auger. Just a few weeks after the Florence meeting, Edoardo Amaldi visited the USA and was excited by a visit to the new Brookhaven Laboratory, where the new Cosmotron was already taking shape. As an advocate of collaborative ventures and as a physicist, Amaldi could not fail to be impressed. Few Europeans had seen a physics effort of such proportions. For his part, Auger in his UNESCO role was working indefatigably and sounded out opinion in many quarters. Over the next two years, his efforts ensured that Rabi's dream became a reality. Having accomplished so much so quickly, it was typical of Auger that he subsequently retired into the background.

Under the auspices of Denis de Rougemont's symbolic but modest European Cultural Centre, a meeting was organized in Geneva, Switzerland, in December 1950 with Auger and delegates representing Belgium, France, Italy, The Netherlands, Norway and Switzerland. Auger unveiled the plan. The proposed new European laboratory would be dedicated to the physics of elementary particles. As an atomic and nuclear physicist, Auger knew that intriguing discoveries had been made in cosmic rays, but that this scanty windfall could become a major harvest, once the big new US accelerators were up and running. The new laboratory would be financed by its member states, Western European nations that were already members of UNESCO (which ruled out Spain and Portugal at the time), but with the possible addition of West Germany (which was not yet a member of UNESCO). Initial contacts with Britain had established that, while British physicists were not against the idea of a European laboratory in principle, they still wanted to go their own way but would be prepared to assist in preparations. At that time, British support was vital to get the idea off the ground, as in Europe only Britain had experience in major projects. Some ideas for proposed contribution levels were discussed. France was initially envisaged as the largest single contributor, with an official figure of 30% having already been tentatively suggested in government circles, and with Germany and Italy each contributing 12.5%. Britain was seen as a minor contributor. For an idea of the overall budget envelope, the meeting looked towards Brookhaven, where US$10 million per year was being spent. Possible sites mentioned for the new European laboratory included Geneva, Copenhagen and Basle-Mulhouse.

The resolution passed at the European Cultural Centre meeting was startling in its ambition or naiveté—depending on the point of view of who read it. It recommended the creation of a laboratory for the construction of a particle accelerator whose power should exceed that foreseen for those currently under construction (which meant the mighty 3 GeV machine which Amaldi had seen being built at Brookhaven and the even bigger 6 GeV machine for Berkeley). In a sector where Europe had no tradition and little expertise, and where no prior in-depth feasibility studies had been done, the idea was simply to jump into the lead by cash and enthusiasm. These bold proposals were initially enthusiastically endorsed in Italy, because of the sterling efforts of Gustavo Colonnetti, President of the Italian National Research Council, and physicist Bruno Ferretti. They were also well received in Belgium, France, Norway, Sweden and Switzerland. In the UK, where physicists were busy building several new machines, there was astonishment and scepticism. 'Who is behind the scheme? Is it serious?' thundered Blackett.

The immediate task was to set up a study group to define the nature of the new particle accelerator project. Candidate names included Cornelis Bakker from The Netherlands, who had built a synchrocyclotron at Amsterdam, Odd Dahl from Norway, a talented engineer responsible for the first nuclear reactor to be built outside the original nuclear 'club' of nations, and the Swiss physicist Peter Preiswerk. For the UK, Cockcroft, much less cynical than Blackett, was in favour of sending a few consultants, such as himself or Chadwick, to spend a few weeks in Paris. In France, Edouard Regenstreif, a specialist in electron optics, was recruited to lead the ongoing technical studies. Among those attending a technical meeting in Paris from 25 April to 3 May 1951 were Odd Dahl and Frank Goward. In a flush of modesty after the strident proclamation after the Geneva meeting, the new advertised goal was to copy the 6 GeV machine proposed at Berkeley, rather than the original idea of the world's largest machine, but this reformulation of the major objective did not satisfy everyone present. Hannes Alfvèn, representing Sweden, was disappointed by the lack of detailed technical information available. However during 1951 support for the new project gained ground, but big questions remained. Where would the new laboratory be built? Who would lead it? Would there be enough money?

Meanwhile a cloud of scepticism hung over northern Europe. The distinguished Dutch physicist Hendrick Kramers, who had participated in the Shelter Island meeting in New York in 1947, suggested to Niels Bohr in Copenhagen that the existing institute in the Danish capital could be used as a centre for the proposed new European institute, an idea which also found favour with Cockcroft and stimulated interest in Norway and Sweden. As President of the International Union of Pure and Applied Physics at the time, Kramers was particularly influential, although physically he was declining (he died in 1952). Bohr, who had not been party to the previous discussions around the Auger–Amaldi axis, began to exert his considerable weight, while others thought that both Bohr and his institute were past their prime and that the new initiative was justified in looking elsewhere. The remoteness and other disadvantages of Copenhagen, such as language, were also a factor. The sheer audacity of going straight for the world's largest machine remained an obstacle. To pacify those who were understandably unwilling to commit themselves immediately to the single lofty goal of a world-class machine, a new plan proposed to build a smaller machine first which could get the new laboratory off the ground. Also at this time, West Germany became a member of UNESCO, opening the door to that country's involvement in the new project.

At a second meeting of technical consultants in Paris in October 1951, proposals were put forward for a 500 MeV synchrocyclotron and a 5 GeV proton synchrotron, with initial design and construction work proceeding in parallel. Four working groups were proposed, one for each of the proposed machines, one for an institute of advanced studies, and a fourth, under Kowarski, for the actual infrastructure of the new laboratory, all under a management board led by Edoardo Amaldi. Delegates seemed optimistic about government funding to the required tune of US$200000, although no firm commitments were yet forthcoming. The meeting also skirted round the question of the site. New criteria, designed to undermine the push to promote Copenhagen, included the use of a well known language, while the possibility of Geneva was continually mentioned and drew little criticism. Overall, solidarity had increased, particularly in the British camp, but Bohr and Kramers still had reservations.

A third meeting of consultants in Paris one month later ventured to suggest that the energy goal of the new synchrotron could be as high as 10 GeV but, to defuse the thorny issue of the ultimate choice of site, separate interim locations for ongoing work were set up around the respective group leaders: Norway and Dahl for the synchrotron, The Netherlands and Bakker for the synchrocyclotron, Kowarski in France for infrastructure, and Copenhagen and Bohr for theory, with Amaldi's administrative hub in Rome.

An intergovernmental meeting at Paris in December under the presidency of the visionary French diplomat François de Rose brought together representatives from 13 European states. For the first time, Germany, with Werner Heisenberg and Alexander Hocker, was party to these open discussions. After continual reiteration of familiar opposing viewpoints, The Netherlands delegation put forward a fivefold plan, with two points designed to appeal to the Northern faction, expressing possible interest in using the existing Copenhagen centre and the Liverpool accelerators, and the remaining points designed to appeal to Franco-Italian sentiment, covering the construction of two new machines and the establishment of 'Conseil Européen pour la Recherche Nucléaire'. The acronym CERN was born.

The Council provided a forum in which developments could be discussed, but a Council cannot do physics. The meeting was continued in Geneva in February 1952, where the provisional CERN Council was voted more specific powers including 'to prepare plans for a laboratory'. The new proposal was immediately accepted by Germany, The Netherlands and Yugoslavia and accepted subject to ratification by Belgium, Denmark, France, Greece, Italy, Norway, Sweden and Switzerland. Together, these nations voted to provide US$211 000, overshooting the initially proposed target of US$200 000, while Denmark offered the new Council the use of premises at the Institute of Theoretical Physics of the University of Copenhagen. While this was a major step forward for CERN, the enigmatic British were not even present at that meeting. It was not the last time that CERN and the UK found themselves out of step. Possibly because the island nation is more prone to extra-continental influences than its continental partners, the UK has on several occasions adopted an iconoclastic role with regard to CERN. However, the first phase of setting up a new European laboratory was over. Emphasis now switched from negotiation and discussion to organization and planning. With the new CERN now open for business, the word 'Council' was no longer appropriate. However, nobody could think of a better acronym, especially when it had to have multilingual appeal, and CERN has stuck ever since. This 'Council', with representatives from each member state, remains CERN's government, normally meeting in Geneva twice a year, but can be specially convened when important decisions have to be made. Slowly the new CERN came to life, blinking in the glare of its new independence, but with the four groups widely dispersed and with Amaldi still in Rome and, when an international telephone call was still a major event, for a while the separate activities became largely autonomous.

A major international physics meeting at Bohr's institute in Copenhagen in June 1952 heard that the first beams had been produced by the new Brookhaven Cosmotron. With this giant machine having been commissioned apparently so easily, confidence in the new technology was high. A CERN Council meeting immediately after the conference advocated that Dahl's group aim for a scaled-up Cosmotron to operate in the energy range 10–20 GeV. Dahl's group, centred in Bergen, now also included a young Norwegian, Kjell Johnsen, while Goward in Harwell acted as Dahl's formal deputy. Despite only a modest formal education, the versatile Dahl, born in 1898, had had a remarkable career. At the age of 24 he was chosen as aeroplane pilot for a polar expedition led by Amundsen. Trying to take off from the ice, Dahl's plane broke up and instead of piloting he spent several years on the ice learning how to make detailed geophysical observations. In 1926 he went to the Carnegie Institute, Washington, and worked with Merle Tuve on one of the early Van de Graaff generators. After the war, his work on a Norwegian–Dutch project for a nuclear reactor introduced him to European collaboration.

If Dahl's group were going to make a scaled-up version of the Cosmotron, it was important for them to go to Brookhaven to admire that machine. In August 1952, Regenstreif, Dahl and Goward made the trip to Long Island. Also passing through was accelerator pioneer Rolf Wideröe, then working on betatrons at Brown Boveri, Zurich, and who was returning from a trip to Australia. To receive their European visitors, Stanley Livingston at Brookhaven had organized a think tank. The Cosmotron's C-shaped magnets all faced outwards, making it

Figure 19: *In August 1952, the visit of a group from the future CERN European Laboratory to the Brookhaven Laboratory near New York indirectly catalysed the development of a new technique which revolutionized the design and construction of big particle accelerators. Left to right, George Collins, Chairman of the Brookhaven Department in charge of the Cosmotron, then the world's most powerful particle accelerator, with three European visitors, namely Odd Dahl, in charge of CERN's project for a major particle accelerator, accelerator pioneer Rolf Wideröe of Brown Boveri, and Frank Goward of CERN.*

Figure 20: *The invention of strong focusing at Brookhaven in 1952 meant that synchrotron magnets could be made much smaller. Left to right, Ed Courant, Stan Livingston, Hartland Snyder and John Blewett. Livingston is holding the equivalent strong focusing magnet against a replica of a magnet from the Cosmotron, which pre-dated the strong-focusing discovery.*

easy for negatively charged particles to be extracted, but not positive ones. 'Why not have the magnets alternately facing inwards and outwards?' suggested Livingston. Ernest Courant, Hartland Snyder and John Blewett seized on the suggestion and quickly realized that this increased the focusing power of the synchrotron magnets. It looked as though the new suggestion, called 'strong focusing' or alternatively 'alternating gradient', because of the arrangement of the magnets, would allow the proton beam to be squeezed into a pipe a few centimetres across, compared with the 20 cm × 60 cm of the Cosmotron beam pipe. The cost of the magnet, the single most expensive item in synchrotron construction, which had to enclose the beam pipe would be greatly reduced, so that a much more powerful machine could be built for the same price tag.

The European visitors arrived at Brookhaven prepared to learn how to make a replica of the Cosmotron and instead learned that the Cosmotron had suddenly become out of date. This visit set the tone for the continuing relationship between the new European generation of physicists and their American counterparts. It was a relationship based on mutual respect, coloured by a healthy spirit of competition, and was to work to their mutual advantage. Untempered, competition can lead to jealousy and secrecy, but in particle physics this has rarely been the case. Although each side has striven to push its own pet projects, collaboration and assistance have always been available, and the community as a whole welcomes and admires breakthroughs and developments, wherever they are made and whoever makes them.

Suitably impressed by what he had seen and learnt at Brookhaven, on his return to Europe Dahl stressed that the new strong-focusing technique had to be used for the new CERN machine. It would gain energy, opening up the prospect of at least 20 GeV, possibly going as high as 30 GeV, and it would save money. The only problem was that nobody had built one yet, anywhere. Undeterred, Dahl, with uncannily accurate judgment, kept his sights firmly on this new goal.

Although a gamble, to take the unexplored strong focusing route turned out to be one of the most influential operational decisions in the history of CERN. It meant that the initial work of the synchrotron group was much more difficult but, had Dahl and CERN played safe, the outcome would have been very different. Brookhaven naturally adopted the new strong-focusing ideas into their repertoire, and Dahl followed his uncanny 'nose', but this was not the case elsewhere. The British, still committed to a national project, had meanwhile set up their National Institute for Research in Nuclear Science near Harwell and were building a 7 GeV traditional weak-focusing proton synchrotron. Other big machines on the drawing boards at Argonne and at Dubna, near Moscow, retained weak focusing. Almost half a century after the invention of strong focusing, these final examples of weak focusing synchrotrons have been left behind. The British and Argonne machines have been closed as high energy machines and the facilities adapted for high intensity machines to produce neutrons instead, while the giant Dubna machine is still operational but is almost a museum piece. However, the prototype strong-focusing machines at CERN and Brookhaven launched following that epic meeting in 1952 are still pumping out beams and play essential roles in the complex of interconnected particle beam supply systems at their respective laboratories.

In Britain, the advent of the strong-focusing proposal gave a new appeal to the European project. The traditional national approach and the more ambitious international venture were no longer directly in competition and became complementary. In such a no-lose situation, resistance to joining CERN began to melt. A draft of the CERN Convention, the organization's constitutional document, had been tabled in October 1952 and was being extensively examined. However before Britain could be persuaded to join the CERN club, a key figure who had to be convinced was Lord Cherwell, formerly Frederick Lindemann, Winston Churchill's staunch friend and advisor. Born in the UK to parents who had left Alsace after it was ceded from France to Germany, Lindemann was educated in Germany, going on to do research with Walther Nernst. After serving as a pilot in the First World War, he had gone on to become Professor of Physics (or experimental philosophy as it was then called) at Oxford in 1919. A contemporary of Henry Tizard, he was greatly offended in 1935 to discover that Tizard had pre-empted a move to set up a national committee for air defence. Although this famous committee soon spawned radar and was thereby assured of success, Lindemann and Tizard went their own ways, each continuing to be highly influential, and giving a curious ambivalence to UK science policy, which continued in the post-war period.

In December 1952, Edoardo Amaldi went on a mission to London, where he was given a frosty reception by Cherwell. Within ten minutes, Cherwell told Amaldi in no uncertain terms that he was sceptical of the CERN idea, while Amaldi said it would be a pity if Britain lost this opportunity. Concluding the brief encounter, Cherwell said the government would duly consider the motion. Undeterred, Amaldi had also told Cockcroft he would like to meet some of the young Britons who might be interested in joining Dahl's group. He met one, John Adams, over lunch at the Savile club. As the rest of the team was at Harwell, for Amaldi to make more contacts meant driving from central London to the former wartime airfield near Oxford which now housed the UK's nuclear research programme. Delegated to act as driver was Adams, the young engineer who had moved to synchrotron development at Harwell after wartime radar work at Malvern. During the trip, a distinct empathy was established between Amaldi and Adams. At Harwell, Amaldi met others who were working on both the CERN machine design and the new machine to be built alongside Harwell, including Mervyn Hine, who had been a student at the Cavendish Laboratory, and John Lawson, formerly of TRE Malvern. While the Harwell machine was committed to the old weak-focusing design, Adams, Hine and Lawson had been taking a close look at strong focusing and had soon discovered that the initial idea was optimistic. Small errors in the magnets (tiny misalignments and field variations) would be naturally amplified and might blow up the carefully accelerated beam. For a time it looked as though this effect might sabotage the scheme. To allow for these effects, the aperture of the new strong focusing machines had to be much larger than had been first been envisaged. A tiny 1 inch aperture was no longer feasible and, to accommodate the larger tube, the enveloping magnet had to be much bigger, 4000 tonnes instead of 800 tonnes, but still considerably smaller than the mighty Russian weak-focusing machine being built at Dubna, with its 36 000 tonne magnet. However, the larger magnet meant that the CERN design would be correspondingly more expensive, and the design energy was compromised to 25 GeV, calling for a 3300 tonne magnet.

Figure 21: *1952 letter to Isidor Rabi from the signatories to the initial agreement to establish CERN. It reads, 'We have just signed the Agreement which constitutes the official birth of the project you fathered at Florence. Mother and child are doing well, and the Doctors send you their greetings.'*

In parallel with these technical advances, the British suddenly switched from aloofness to commitment and formally joined CERN, while the CERN Council finally decided the site for the new laboratory. Copenhagen was too far north and too dominated by Bohr, Paris was too dominated by the French, and the strongly supported case for Arnhem in Holland was not supported by the French. This left Geneva in Switzerland as the surviving option. On the map, the canton of Geneva appears a curious appendage at the extreme west of Switzerland. Almost totally surrounded by France, it is joined to the rest of

Switzerland by an umbilical cord only a few kilometres across. This geographical peculiarity has been emphasized by the establishment there of many international organizations, beginning with the Red Cross in 1863. 1936 saw the opening of the grand 'Palais des Nations', the headquarters of the League of Nations, and which in 1946 became the European headquarters of the new UN. Geneva had thrived on internationalism, but opposition to the proposal to offer CERN a home came from an expected quarter, an element of the local community, who, misunderstanding the role that CERN would play, feared radiation problems. A referendum in Geneva on 27–28 June 1953 defeated the anti-CERN initiative, and the way was clear. However, the 'N for nuclear' in CERN's title has dogged it ever since. In the 1980s, the laboratory's title was changed to 'European Laboratory for Particle Physics', but the handy CERN acronym remained and, with most of the local population still unsure of the exact business of the huge laboratory in their midst, occasional (and unwarranted) 'N for nuclear' scares have continued.

With the green light from Switzerland definitive, an advance party of the proton synchrotron group arrived in Geneva and were joined by John and Hildred Blewett, seconded from Brookhaven. However, Dahl preferred to remain in Bergen, appointing Goward as on-site supervisor. Dahl's decision to go for the new strong focusing option was very unselfish. 'Instead of being an engineering group scaling up an existing machine based on well-established principles, it became a physics group studying the theory of accelerators, only returning later to engineering design,' said Kjell Johnsen later. Realizing he could not handle this commitment, Dahl preferred to stay in Bergen and resigned. Just a few months later, in March 1954 and aged only 33, Goward collapsed and subsequently died of an unsuspected brain tumour. John Adams, at the tender age of 34, became leader of the proton synchrotron project. After Dahl's decision to back the dark horse of strong focusing, fate provided the project with the right leader. John Adams was to be the Moses who would take CERN to the Promised Land.

Having made so many wise decisions, it was inevitable that there had to be a bad one. In 1953, the CERN Council set up a committee to select and appoint a Director General to head the new laboratory. Surprisingly, neither of the two founding fathers, Auger and Amaldi, were interested. Other candidates were Hendrik Casimir, the distinguished physicist who had become a director of the Philips industrial giant, and Blackett. By April 1954 the choice had settled on Felix Bloch from Stanford, who had shared with Edward Purcell of Harvard the Nobel physics prize in 1952 for their independent measurements of the magnetism of nuclear particles. The Nobel prize crowned Bloch's career. Born in Zurich in 1905, he studied with Heisenberg in Leipzig in 1928, developing a famous theory of electrons in crystals. After work with Bohr in Copenhagen and another spell in Leipzig, he was forced to leave Germany in 1933, eventually settling at Stanford. There could be no doubt that Bloch was being chosen as a figurehead leader for CERN. A European physicist who had emigrated to the USA returning to the land of his birth (although in the meantime he had taken out US citizenship) was felt to be highly symbolic. With his Nobel prize and long list of physics accomplishments, he would endow the infant laboratory with additional prestige. However, Bloch had not been involved at all in any preparatory CERN work and made it clear that he was reluctant to move, insisting that, if he did so, he would continue his work on nuclear magnetism, bringing his apparatus and research staff with him at CERN expense. The CERN Council agreed to Bloch's demands. Bloch would be aided by Edoardo Amaldi

Figure 22: *At an early CERN Council meeting in 1953. Left to right: Edoardo Amaldi, Jan Bannier representing the Netherlands, and Pierre Auger and Jean Mussard of UNESCO.*

as Deputy Director, with Cornelis Bakker representing the scientific group leaders. Arriving in Geneva in September 1954, Bloch tried to delegate much of the CERN work to Amaldi, who had already turned down the chance of becoming Director General himself and was eager to return to scientific research. The new CERN administration was quickly on the rocks. Bloch tendered his resignation almost immediately and left CERN the following year. Bakker, who had been offered the job of interim director in October 1953 and had turned it down, succeeded him. The machine sector at CERN went on to maintain a tradition which had been stamped by the far-sightedness of Odd Dahl and the technical skills and exigencies of his lieutenants, and the physics research side eventually matured to a similar excellence, but the Bloch episode was not the last time political pressures tried to push square pegs into round management holes.

CERN's first machine, the 600 MeV synchrocyclotron, was commissioned in 1957 and was soon producing its first physics results. By that time, with Brookhaven's Cosmotron and Berkeley's Bevatron in full physics production, and with the big new Dubna machine attaining 10 GeV, the arrival of the CERN synchrocyclotron boosted morale at the new Geneva laboratory. Outgunned from the start even by the Birmingham synchrotron, the new 600 MeV machine was a minor feature on the world physics landscape. Eyes were focused on the race between the proton synchrotron teams at Brookhaven and at CERN.

On 24 November 1959, the evening before Hildred Blewett returned to Brookhaven, CERN's new proton synchrotron unexpectedly accelerated protons all the way to 25 GeV, becoming the world's highest-energy machine, with its strong focusing easily outstripping Dubna's 10 GeV. While the CERN team was jubilant, it was also slightly perplexed at its sudden success. Kjell Johnsen, for one, would have liked the commissioning to have been more sedate, so that they could have celebrated each increase in energy. 'Now there is only one moment of triumph—we have been cheated!' he said. There were emotional and dramatic scenes which offered a sharp contrast in national stereotypes. Gilberto Bernardini from Italy jubilantly kissed a disconcerted John Adams on both cheeks. The laconic Adams phoned up Alec (later Sir Alec) Merrison, who later recalled, 'He did not tell me in highly excited tones. He said, 'Remember those scintillation counters you and Fidecaro put in the ring? Will they detect 20 GeV protons?' I paused long enough to grab a bottle of whisky and Professor Fidecaro, in that order, and came along to celebrate ...' . The following day, at a special meeting to convey the news to CERN staff, John Adams showed a vodka bottle which he had been given some months earlier on a trip to Dubna with strict instructions that it should be drunk when the CERN proton synchrotron surpassed Dubna's 10 GeV energy. The bottle was now empty. Adams produced a photograph of the oscilloscope trace displaying the achievement and squeezed it into the bottle, giving instructions that it should now be returned to Dubna.

However, Dubna's vodka bottle was not the only empty thing at CERN. Also very empty were the experimental halls around the new synchrotron. In the rush to build the new machine, few people had given much attention to the instrumentation needed to carry out experiments. Over at Brookhaven, where Director Leland Haworth had formally proposed a new Brookhaven machine to the AEC in 1954, the new Alternating Gradient Synchrotron (AGS), the 'twin' of the new CERN machine, did not supply beam until six months later, but this delay was more than compensated by the enthusiasm and ingenuity which went into planning experiments. US physicists had cut their high-energy accelerator teeth on the Cosmotron and the Bevatron. Within a few years of its commissioning, the AGS was to reap an impressive harvest of new physics results. At CERN, the cupboard was to remain bare. The new laboratory had risen to the challenge of building the world's most powerful machine from scratch in just a few years, but developing the physics research infrastructure around the machine and fostering experimental prowess was to take much longer.

THE QUARK IN THE BOX

While diplomatic moves focused on setting up the new European laboratory in the early 1950s, the new Brookhaven Cosmotron was alone in a new energy range. Lower-energy synchrocyclotrons had already provided beams of pions, the particle predicted by Yukawa as the carrier of the nuclear force and discovered in cosmic rays at Bristol in 1947. By shining these pion beams on nuclear targets, physicists hoped to discover how the nuclear force operated. Apart from their different electric charges, why were the proton and the neutron so similar? How could many protons and neutrons stick together to form nuclei, overcoming the intense electrical repulsion between closely packed protons?

Physicists thought of the nucleus as a smaller version of the atom. The atom (a central nucleus and orbital electrons) was held together by electromagnetic forces, transmitted by photon messengers which 'told' the nucleus that it had electrons for company. Quantum mechanics had shown that the microworld, even the vacuum itself, is full of spontaneous flashes of energy whose duration is inversely proportional to the energy or 'brightness' of the flash. The brighter the flash, the briefer it is. A single electron, for example, produces a fusillade of photons, electromagnetic flashbulbs which advertise the electron's presence and even affect the properties of the electron itself, as Feynman and Schwinger had so elegantly demonstrated in their new picture of quantum electrodynamics. Photons, being massless, can roam over wide distances, making the electromagnetic force a long-range effect, with long-haul photons operational over much longer distances than atomic dimensions.

The nuclear force, on the other hand, has a limited range. The Yukawa mesons, the supposed carriers of nuclear force, are relatively heavy and cannot roam as widely as photons. With beams of pions, physicists had a key which might open the locked door to the nuclear force. Unfortunately, the nuclear force turned out to be very different from this naive picture of a nuclear 'atom' held together by Yukawa pions. Opening the door to the nuclear force would eventually require the equivalent of a subnuclear oxyacetylene torch rather than a key, but first the pion key had to be tried.

The first experiments with pion beams at the Chicago and Carnegie synchrocyclotrons had suggested that at certain energies, instead of piercing the nuclear target, the pions started to react with nuclei, with widespread fragments emerging. Incident pions made the nucleus 'ring' like a bell, vibrating in all directions. At the higher-energy Cosmotron, these nuclear chimes (resonances of the nuclear protons and neutrons which go off like an alarm when a pion intruder comes near) were heard more distinctly. The lifetime of these nuclear resonances is so short that, even travelling at the speed of light, they hardly have time to move before disintegrating. They are too transient to warrant being called particles in their own right but, by studying this new nucleon spectroscopy, physicists hoped to get a fresh slant on the nuclear force.

Not far behind the Cosmotron was the Bevatron at Berkeley. While the Cosmotron was basically a 'look–see' machine, exploring an uncharted energy domain, the Bevatron was built with a definite mission. When Paul Dirac's 'designer equation' with inbuilt relativity predicted the existence of an anti-electron in 1931, the courageous Dirac also ventured that other particles obeying similar equations should also have antiparticles and predicted the existence of the antiproton, the counterpart of the electrically charged nuclear particle.

The Bevatron's aim was to discover the antiproton, but to do so it would have to

Figure 23: *The Bevatron at Berkeley was built with a definite mission, namely to find the antiproton, the antimatter counterpart of the proton. When it began operation in 1954, it was the world's most powerful proton machine. By the following year, it was delivering proton energies beyond the 6 GeV antiproton production threshold. A team led by Emilio Segrè and Owen Chamberlain soon saw their first eagerly awaited antiprotons. Chamberlain and Segrè went on to share the 1959 Nobel prize.*

attain a new energy horizon. The machine accelerated protons, but these protons could not themselves transform into antiprotons. Matter cannot transform into antimatter. The protons first had to produce other particles, and then these in turn would produce proton–antiproton pairs. The Bevatron had to supply enough energy to make these extra particles. When it began operation in 1954, it was the world's most powerful proton machine. By the following year, it was delivering proton energies beyond the 6 GeV antiproton production threshold. In the space of a few months, a team led by Emilio Segrè and Owen Chamberlain had seen about a hundred of· the eagerly awaited antiprotons, the first examples of a nuclear antiworld. The Bevatron had lived up to its promise, and Chamberlain and Segrè went on to share the 1959 Nobel prize but, in this initial experiment, there were no picture-postcard antiproton tracks. The antiprotons were identified by electronically timing them over a 12 m flight path, the first particle to be discovered by such methods.

This discovery emphasized how the classic method of recording particle tracks, the cloud chamber, was fast running out of steam. The new synchrotrons delivered pulses of particles every few seconds, too fast for the cloud chamber's cumbersome expansion. In addition, penetrating high-energy particles are not easily stopped by gases or vapours. Cloud chambers would record tracks of particles passing through, but not many interactions. In April 1952, an early experiment at the Cosmotron, using an adapted version of the cloud chamber, reported its first examples of V particles, the tracks first seen in the Manchester cosmic ray experiments. These desperate attempts to update the cloud chamber were not enough. The faithful detector technique invented by C T R Wilson at the Cavendish Laboratory half a century before had been squeezed dry. The future lay with electronic techniques, such as had been used to discover the antiproton, or with some new method to reveal high-energy tracks from pulsed machines.

In 1952, Donald Glaser at the University of Michigan had the idea of doing with liquids what the cloud chamber did with gases. He showed that suddenly reducing the pressure in a liquid could make it boil along the track left by a particle. Because a liquid has more stopping power than a gas, it is better suited to recording the tracks of high-energy particles. In 1954 at the Cosmotron, Jack Steinberger, who had participated in the experiment that discovered the electrically neutral pion at the Berkeley synchrocyclotron, mounted the first bubble chamber physics experiment and was rewarded with the lambda, a heavy cousin of the proton. Berkeley had also quickly cottoned on to the new detector technique. Lawrence's laboratory liked to do things in a big way, and Luis Alvarez, who had built Berkeley's big linear accelerator for protons, set out to apply the bubble chamber on a big scale. Just as Lawrence's cyclotrons had quickly grown in size as confidence grew, so Alvarez went in quick succession from an initial 6 cm diameter chamber in 1954 to 8 cm, to 25 cm, to 75 cm, and finally to 180 cm (72 inches) in 1959. The bubble chamber arrived just at the right moment.

With all this input from the Bevatron and the new bubble chambers, by the end of the 1950s the original neat nuclear pair (the proton and the neutron) had grown to become an extended family of several dozen nuclear resonances and unstable particles—too many for comfort. Physicists were reminded of the abundance of chemical elements discovered in the nineteenth century, when chemists had extended their list from a few common substances to dozens. Instead of being overawed by this extravagance, the periodic table had exploited it to reveal underlying regularities. In the same way, the new gamut of nuclear states was to reveal underlying nuclear patterns which had been invisible with the nuclear horizon restricted to the proton and the neutron.

The proton and the neutron are close relatives; both live inside everyday nuclei, where they are both stable. However, if a neutron is ripped from its nuclear home and left to roam free, it is unstable, decaying within about ten minutes to a proton and an electron. It made sense to talk of the proton and the neutron, to a first approximation, as a single entity, the 'nucleon', which came with a heads-or-tails option, differentiated by electric charge. Among the long list of newly discovered states, other examples of such electric charge kinships could be seen. Some, like the proton and neutron, had two members, some had three, and some four. Particles also resembled each other in other ways. Unstable particles decayed into lighter, more familiar particles; the kaon decayed into the pion, and the heavy lambda produced a nucleon. There were regularities in all these decays, certain particles preferring to keep each other company. Physicists spoke of 'associated production'. In 1955, Murray Gell-Mann, working in Chicago, and Kazuhiko Nishijima, in Osaka, put forward a new scheme which suddenly made sense of these rules.

Gell-Mann, like Feynman and Schwinger, was a product of the ebullient metropolis of New York City. After an introduction to physics research at MIT, Gell-Mann went on to join Fermi's group in Chicago. Gell-Mann has an uncanny ability to spot regularities and rules; a lifetime interest has been linguistics, where he is able to discern how obscure languages work. Fly Gell-Mann into an unusual conference venue and he will soon make an incisive remark about the etymology or pronunciation of the language there. Gell-Mann's major contribution to physics came when he saw patterns appearing in the production of the new particles. These regularities could be understood if, as well as their electric charge, the nuclear particles also carry another label, which Gell-Mann called 'strangeness'. Strangeness was the first of several characteristic Gell-Mann contributions to the vocabulary of particle physics. Strangeness was a small integer number, which, like electric charge, had to be conserved in a reaction. It also upset some people. Used to working with

Figure 24: *A mark for Mr Quark. In 1964, Murray Gell-Mann's quark ideas revolutionized particle physics. This photograph, taken by CERN physicist Maurice Jacob in a London pub, shows Gell-Mann with the manuscript of his book* The Quark and the Jaguar.

particles labelled by Greek letters or whose exact meaning could only be understood by scholars, some physicists thought Gell-Mann's new term was an insult to their intellect. Gell-Mann, who is more intellectual than most of his critics, thought otherwise, and the bizarre name stuck. Strangeness turned out to be the key to the plethora of new particles. Gell-Mann, now at Caltech, Pasadena, and the Israeli physicist Yuval Ne'eman, working independently in London, discovered that plotting the electric charge of particles against their strangeness gave neat patterns. These were mainly octets, containing two charge doublets like the proton and the neutron, a charge triplet, and a charge singlet at the centre.

Another octet accommodated lighter nuclear particles, like the pion and the kaon. Bubble chamber studies at Berkeley discovered more such particles, calling for a third octet. However, there was one place where octets did not seem to fit. It was difficult to make an octet with the charge quadruplet (charges 2, 1, 0 and −1) of pion–nucleon resonances. It looked as though these nuclear states had to be fitted into a larger picture but suddenly, instead of there being too many particles, there were not enough. In July 1962 at a physics conference at CERN came the news that a California group had discovered a new heavy resonance, the xi, which came in electrically neutral and negative varieties. In the audience Gell-Mann and Ne'eman immediately realized that here were two more pieces to fit into their jigsaw. The new xi carried strangeness minus two, and the charge–strangeness diagram, instead of being an octet, now resembled a pyramid, with four pion–nucleon resonances at the bottom, three sigma particles in the next row, and the pair of newly discovered xi resonances in the third. Both Gell-Mann and Ne'eman knew that to complete such a pyramid a single particle was needed, with strangeness minus three. With the heavy xi still undiscovered, neither had ventured to predict a further missing particle. The next day, at the

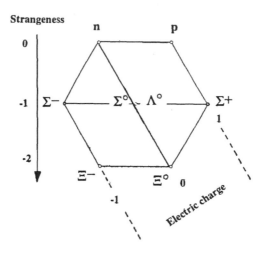

Figure 25: *Plotting the electric charge of particles against strangeness gave neat patterns, like this hexagonal octet containing two charge doublets (the proton (p) with the neutron (n), and the neutral xi (Ξ^0) and negative xi (Ξ^-)), a charge triplet of three sigma particles (Σ^+, Σ^0 and Σ^-) and the singlet lambda (Λ).*

end of a review talk on new particles, both Gell-Mann and Ne'eman raised their hands, but the more well known Gell-Mann was recognized first and invited to speak. He strode to the blackboard and filled in the missing tenth particle at the apex of the pyramid, calling it the omega, with a minus qualifier, as it had to have negative electric charge. From the regularities in the masses of the other decuplet particles, he also predicted the omega's mass.

In the audience was Nicholas Samios, formerly a student of Jack Steinberger who had participated in the first Cosmotron bubble chamber experiment, and now in a group which was preparing a study using a new 80 inch bubble chamber at the Brookhaven AGS. The Brookhaven bubble chamber group also included Ralph Shutt, who had participated in the cloud chamber experiment at the Cosmotron, the first time that V particles had been seen at an accelerator. At the CERN meeting, Gell-Mann emphasized to Samios the importance of using a beam of kaons to supply enough strangeness to make the missing omega.

The experiment started to run just before Christmas 1963 and, by the end of January, had accumulated 50 000 bubble chamber photographs, all of which were vigilantly scanned for signs of the omega. On 31 January 1964, Samios' team found it. It broke the temporary monopoly of particle discoveries by Alvarez' bubble chambers at Berkeley and showed that the Gell-Mann–Ne'eman classification scheme was right. Brookhaven had not been the only laboratory looking for the omega minus. After all the particle had been publicly predicted at CERN. Early in 1964, a CERN bubble chamber revealed something that looked like the decay of an omega minus but, as some of the decay products had escaped the bubble chamber undetected, no sure identification could be made. European physicists had to swallow their pride and admire Samios' picture postcard example with even invisible particles choosing to produce charged-particle pairs right there in the bubble chamber. As British theorist Paul Matthews wryly remarked in his contemporary book *The Nuclear Apple*, 'God is an American'.

What lay behind octets and decuplets of particles proposed by Gell-Mann and Ne'eman? Physicists realized that such patterns were the result of an underlying threefold symmetry. In Japan, Shoichi Sakata had suggested that these patterns were derived from an underlying triplet of particles: the proton, the neutron and the strange lambda. However, Gell-Mann realized that the underlying threefold symmetry had to involve a different pattern of electric charge from that proposed by Sakata. The problem was that these electric charges were fractional, either two-thirds or one-third of the charge on the electron, and nobody had ever seen such fractional electric charges. There was no smaller visible quantum of electricity than a single electron. During a trip to New York's Columbia University in 1963, Gell-Mann developed a picture which was to change the face of physics, but first it needed a name, and Gell-Mann enjoyed names. To himself he called his fractionally-charged particles 'quorks', like a play on 'quirks' of his imagination. A favourite book of Gell-Mann was James Joyce's enigmatic *Finnegan's Wake*, with its famous phrase 'Three quarks for Muster Mark', in which the word 'quark' looks as though it rhymes with 'Mark'. However, Gell-Mann is convinced that Joyce's 'quark' is meant instead to resemble 'quart', so resembling his original 'quork' sound. Most people pronounce 'quark' as though it rhymes with 'mark' and 'bark', except Gell-Mann, who carefully adheres to his original 'quork' pronunciation. However it is pronounced, it became Gell-Mann's second major contribution to the vocabulary of modern physics and lived up to the quirkiness of 'strangeness'. Remembering the furore which his proposal of strangeness had stirred up, Gell-Mann sent his first quark paper for publication in a European journal, *Physics Letters*, instead of the 'pompous' but more prestigious US journal *Physical*

Review Letters. It came out in print at the same time as the news of Samios' omega minus at Brookhaven. Quarks had arrived, and physics seminars began to sound like Donald Duck cartoons.

However it was pronounced, the quark picture was too compelling to disregard. Suddenly everything made sense. According to Gell-Mann, the heavy nuclear particles were built up of three quarks: one, called 'up', carrying two thirds of a unit of positive electric charge, a second, called 'down', carrying a third of a unit of negative electric charge, and a third, called 'strange', also carrying a third of a unit of negative electric charge. In addition the strange quark has one unit of Gell-Mann's strangeness. A proton is made up of two 'ups' and a 'down', while a neutron had one 'up' and two 'down'. The new omega minus was made up of three strange quarks. Complementing the quarks was a mirror world of quark antimatter: antiquarks.

The fact that nobody had ever seen such proton or neutron constituents or

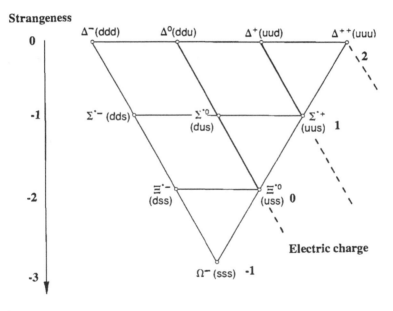

Figure 26: *The tenfold 'decuplet' of nuclear particles and their composition in terms of 'up' quarks (u, carrying electric charge ⅔), 'down' quarks (d, carrying charge −⅓), and strange quarks (s, carrying charge −⅓ and strangeness −1). At successive levels of strangeness, first a charge quartet (Δ⁺⁺, Δ⁺, Δ⁰ and Δ⁻), then a charge triplet (Σ*⁺, Σ*⁰ and Σ*⁻), followed by a charge doublet (Ξ*⁰ and Ξ*⁻), and finally the famous omega minus (Ω⁻), composed of three strange quarks.*

fractional electric charge did not bother Gell-Mann unduly. He concluded that quarks are not free particles which can be knocked out of their natural habitat in the same way that protons or neutrons can be dislodged from nuclei or that electrons are ripped from their atomic orbits. Quarks are there but are permanently confined in some way. Gell-Mann called his quarks 'mathematical' particles, rather than 'real' particles. In his 1964 quark paper, Gell-Mann said, 'A search for stable quarks ... at the highest-energy accelerators would help to reassure us of the non-existence of real quarks.' This was subsequently interpreted by many as implying that Gell-Mann did not believe quarks were there at all and were just some mathematical device to make everything come out right, an interpretation that Gell-Mann rejects.

All nuclear particles now had a quark assignment. The proton and the neutron and their heavier cousins were made up of three quarks. The pion and the kaon and all their new relatives which had been discovered in Berkeley bubble chamber experiments could be explained as a quark and an antiquark locked together. Moreover, simple quark calculations gave good predictions, and Gell-Mann's prediction of the omega minus mass at the 1962 CERN physics conference now had a powerful underlying explanation. Each time that an extra strange quark was added, the particle mass increased by 150 MeV. The idea of quarks was so compelling that physicists immediately began to look for them, peering in all sorts of places: in high-energy experiments, in cosmic rays, in oysters, in Moon rocks, and in precision reruns of the experiment which had originally measured the charge on the electron. Although from time to time there was a shout from one or other of these searches, no sign of free quarks has been found, but Gell-Mann's 'mathematical' quarks had to be there.

The bubble chamber experiment which discovered the omega minus had swung the physics pendulum from the US West Coast to the East Coast. The discovery that things such as quarks were really there was to make it swing back again. In the 1950s, Robert Hofstadter at Stanford used the Mark III electron linear accelerator to take an electron 'x-ray' of protons. Electrons, being very small, can pierce deep inside matter and reveal tiny details of electrical structure. Hofstadter's experiments had shown that the proton is not a point but has a shape, a fuzzy blob a hundred thousand times smaller than the atom, a new insight which merited the 1961 Nobel Prize. Hofstadter, together with his Stanford colleague Wolfgang ('Pief') Panofsky, realized that higher-energy electron beams would be able to see more clearly what a proton looked like. In 1962 groundbreaking began for a monster 2 mile machine which slashed across the landscape near Palo Alto. A bridge midway along the vast machine anticipated the construction of a new highway. Although spawned by Stanford University, the big new machine became a national laboratory in its own right, the Stanford Linear Accelerator Center (SLAC). Its first director was Panofsky, the son of a German art historian who had joined the pre-war mass emigration.

In 1967, the new SLAC machine began to fire 18 GeV electron beams at targets. If there was anything to be found inside the proton, this would find it, and to accomplish this it had been fitted with giant 'eyes', one of which had been built by Jerome Friedman and Henry Kendall, who had earlier worked on the smaller Stanford linear electron machines and then migrated to MIT, and by Richard Taylor, who had joined the new SLAC.

At first, their big detector saw the same sort of proton fuzz that Hofstadter and other people had already seen but, with the detector away from the line of the incoming electrons, it began to see something else. There was a steady signal of collision fragments coming out at these larger angles. It was a higher-energy rerun of the classic 1909 alpha particle-experiment by Ernest Rutherford, Hans Geiger and Ernest Marsden which had discovered that a tiny hard nucleus lived at the centre of the atom. Over half a century later, Friedman, Kendall and Taylor found that most of the time the proton is a fuzzy blob through which the SLAC electrons could slice like a hot knife through butter but, occasionally the electrons hit something very small and very hard from which they ricocheted violently. Friedman, Kendall and Taylor were awarded the 1990 Nobel prize for their discovery of a deeper layer in the structure of matter.

Instead of track photographs, the big SLAC detector gave momentum distributions and the like. Understanding these results needed insight into subnuclear kinematics. James Bjorken, a SLAC theorist, had prepared the ground. Just as Rutherford's discovery of the nuclear atom had put paid to the old picture of a fairly amorphous atomic 'pudding', so in the same way, Bjorken had suggested, the nucleus might also have a deeper layer of constituents with a very different structure. Bjorken had shown how this deeper layer might be revealed in experiments. Richard Feynman, visiting SLAC as the new results began to emerge, interpreted Bjorken's picture and the new measurements in terms of tiny electrically charged particles (called by Feynman 'partons') roaming around at will deep inside a rigid proton box. It was tempting to identify these subprotonic partons with Gell-Mann's quarks. There were indeed hints that the tiny hard parton grains inside the proton behaved as though they had quarks' quirky fractional charge, but the SLAC momentum distributions showed that these partons seemed to be moving freely inside their proton box. How could freely moving partons be reconciled with Gell-Mann's quarks which had to be firmly locked inside nucleons? In addition, the experiments measured the fraction of the proton's energy which was carried by the parton. If the proton were simply a box of three quarks, each quark would contribute about one third of the proton's energy. However analysing the SLAC results showed that the struck parton could carry a wide range of fractional proton energies. The SLAC electrons were sensitive to all electrically charged particles, but adding up the fractional energies of all partons came to only about one half; only 50% of the proton's energy is carried by electrically charged particles. Where did the rest of the proton's energy go? Whatever it was, it had to be electrically neutral; otherwise it would have been discerned by the electron beams. To find out more about partons required another sort of beam.

Quarks were introduced by Murray Gell-Mann to explain the proliferation of nuclear particles, but most nuclear particles are unstable. If a nuclear particle, composed of quarks, is unstable, then its quarks have to be unstable too, with one quark changing into another, giving a new quark configuration. What catalyses such a quark transition? How can such transitions be studied? While the electron beam experiments at SLAC had 'x-rayed' the proton and seen its quarks, electrons could only differentiate quarks by their electric charge. However, there was another kind of beam which as well as seeing the quarks could transform one quark into another. The development of this technique has itself been one of the great scientific sagas of the twentieth century.

At the turn of the century, physicists discovered that certain nuclei underwent 'beta decay', changing into other nuclei and giving out electrons. For the next thirty years, these beta-decay electrons were the subject of intense study. While other subatomic phenomena gave a ladder-like set of discrete spectral lines, reflecting quantum jumps, the energy distribution of beta-decay electrons was smooth, and there was another problem. In any nuclear decay, the energy of the parent nucleus has to match the sum of the energies of the decay products. Relativity had shown how mass (rest energy) had to be taken into account in this energy accounting. In contrast, in beta decay, most of the time the energy of the parent nucleus was greater than the sum of the energies of the daughter nucleus and the emerging electron.

To make the energies balance, something else had to emerge from the nuclear disintegration as well. Try as they might, the physicists could not see anything. Other than the energy mismatch between the product nucleus and its accompanying electron, nobody could find any sign of an accompanying particle. This 'missing energy' was boldly predicted in 1929 by Wolfgang Pauli to be an invisible 'ghost' particle, eventually called by Fermi the 'neutrino' (the neutral little one). In the 1930s, physicists resigned themselves to the fact that neutrinos were arrogant particles which simply ignored matter. Their only role was to carry off energy in beta decay. Otherwise they did nothing. After a series of diligent neutrino searches drew a blank, calculations showed that a single neutrino had a good chance of passing through the Earth without being captured.

In those early days, Pauli, Fermi, Bethe and Peierls and the other physicists who had thought about the neutrino had not realized that one day the products of nuclear fission in reactors would be pumping out beta decays at an unprecedented rate (ten million million neutrinos per square centimetre per second). If the normally negligible chance of intercepting a single neutrino were multiplied by such large numbers, the result was no longer negligible and neutrino detection could become feasible. In 1956, Frederick Reines and Clyde Cowan of Los Alamos sent a telegram to Pauli in Zurich informing him that they had detected the ghostly neutrinos, to which Pauli replied, 'Thanks for the message. Everything comes to him who knows how to wait.'

Having discovered neutrinos, the next challenge was to use them. Bruno Pontecorvo, a flamboyant Italian physicist, had many ideas for neutrinos, even before the particles had been discovered. Pontecorvo had worked with Fermi in Rome and then moved in turn to Paris and the USA before joining the British nuclear research programme, first in Canada and then at Harwell. In 1950 he suddenly disappeared, causing political waves by turning up several years later in Russia. With neutrinos discovered, Pontecorvo said the particles should be put to good use, but exploiting neutrinos called for ingenuity on a new scale.

Jack Steinberger and Leon Lederman first met at Columbia University, New York, in 1951 when its 385 MeV synchrocyclotron, then the world's highest-energy accelerator, was starting up. Mel Schwartz, who had once been Steinberger's student at Columbia, had returned there after a stint at Brookhaven. In 1958 at Columbia two things were in the air: neutrinos had just been discovered, and the new AGS was nearing completion nearby at Brookhaven. To inaugurate the new machine, what better than a pioneer experiment using neutrino beams? This posed two major challenges: a considerable amount of shielding, to block off all other particles and to leave only the neutrinos, and a big detector to be sure of catching some of the elusive particles. The trio used a new type of detector, a spark chamber, where particle tracks showed up as a series of electrical discharges between multiple layers of high voltage plates and could be photographed.

Designing neutrino detectors is a delicate task. Make them too small and they catch no neutrinos. Make them too large and they are too expensive and unwieldy. Lederman, Steinberger and Schwartz' spark chamber weighed 10 tonnes. These man-made neutrinos, accompanied by muons, came from the decay of pions and kaons, and the 1962 experiment at Brookhaven revealed that these synthetic neutrinos behaved in a different way from the neutrinos which had been seen coming from nuclear fission. The experiment showed that neutrinos, which undergo remarkably little reaction anyway, moreover come in two kinds: one associated with electrons and the other with muons. The discovery earned a Nobel prize for the American trio, although they had to wait until 1988 for their trip to Stockholm.

CERN had also hoped for a neutrino beam at its new proton synchrotron but initially had got its calculations wrong and had underestimated the number of parent particles needed to create a useful beam. CERN had hastily to reappraise

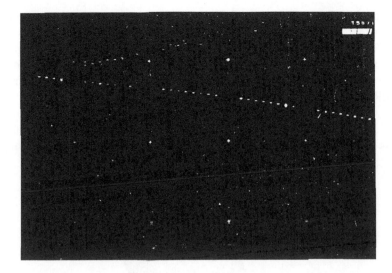

Figure 27: *First tracks from synthetic neutrinos. In 1962 an experiment at Brookhaven revealed that these synthetic neutrinos behaved in a different way from neutrinos from nuclear fission. The experiment used a new type of detector, a spark chamber, where particle tracks showed up as a series of electrical discharges between multiple layers of high-voltage plates.*

its neutrino programme and had watched jealously from the wings at Brookhaven's solo performance in the first act of the neutrino play. Dejected European physicists hearing the Brookhaven news at a physics meeting at Sienna, Italy, in 1963 realized that they would have to be more ambitious if they wanted to get in on the neutrino act. To compensate for the initial underestimation of beam strength, the Dutch engineer Simon van der Meer at CERN had the bright idea of using a giant magnetic 'horn' to focus the particles before they decayed, thus boosting the neutrino intensity. This alone was not enough. Just as the Brookhaven debut neutrino experiment had introduced a detector innovation, then CERN too had to find a detector to match the neutrinos' elusiveness.

To increase the chances of catching the elusive neutrinos, bubble chambers were filled with heavy liquids such as propane or Freon instead of hydrogen, but this meant that the interactions would become much more complex. Luis Alvarez likened heavy-liquid bubble chambers to catching particles in 'strawberry jam'. However, CERN's initial modest neutrino experiments with a bubble chamber 1 m long saw that neutrinos were less elusive at higher energies. Some far-sighted physicists were intrigued by this phenomenon, but at the time there had been no explanations on the market. The observation was filed away and gathered dust.

In parallel, CERN built a 1 GeV 'Booster' machine to increase the supply of protons in its synchrotron. Coming into action in 1972 for the second generation of CERN neutrino studies, the Booster increased the neutrino yield. Dwarfed by its giant neighbour synchrotrons, the Booster is easily overlooked but, as the first link in a chain of interconnected machines, has gone on to play an important role in CERN's particle beam schemes.

For a second-generation detector at CERN, André Lagarrigue, who had learnt physics in Leprince-Ringuet's cosmic-ray laboratory, had the idea of a mighty

Figure 28: *André Lagarrigue, who had learnt physics in Leprince-Ringuet's cosmic-ray laboratory, had the idea of a mighty bubble chamber containing 18 tonnes of heavy liquid (Freon) to intercept CERN's neutrino beams. Called Gargamelle, after the mother of Rabelais' giant Gargantua, the huge bubble chamber, weighing in total 1000 tonnes, was built by the Commisariat à l'Energie Atomique and came into action in 1970.*

bubble chamber containing 18 tonnes of heavy liquid (Freon) to intercept the CERN neutrinos. Lagarrigue had initially planned a small 1 m heavy-liquid bubble chamber for the initial round of neutrino studies, but this had fallen victim to the last minute reappraisal. For the next round, Lagarrigue set his sights high. Called Gargamelle, after the mother of Rabelais' giant Gargantua, the huge bubble chamber, weighing in total 1000 tonnes, was built by the French Atomic Energy Commission and came into action in 1970. Before Gargamelle, intercepting neutrinos had called for physics sleuth work but, with Gargamelle, the neutrinos had met their match. Ready to analyse the photographs was a large European collaboration of physicists from Aachen, Brussels, London, Milan, Oxford and Paris, as well as CERN. The scanning tables alone for measuring and analysing the bubble chamber photographs cost the equivalent of US$0.5 million each.

The idea was now to use neutrinos to 'x-ray' the deep interior of nucleons. It was a rerun of the 1967 SLAC experiment that had discovered how the proton behaves as a box of tantalizingly quark-like partons. However, electrons interact with partons only through their electric charge and are blind to the specific labels that the quarks carry. Neutrinos, on the other hand, can differentiate between one quark type and another.

A simple example of nuclear beta decay is when a free neutron decays into a proton. In quark terms, this means that one of the two 'down' quarks in the neutron transforms into an 'up' quark, spitting out a neutrino and an electron as it does so. Nuclear beta decay is the reflection of a quark transition deep inside one of the nuclear particles. The idea with neutrino beams is to shuffle such a quark reaction around, using the neutrino to switch one quark into another. Although making new particles in high-energy collisions was not new, neutrino beams provided a new precision scalpel which reached to the quark core of nuclear particles. For example a neutrino hitting a down quark switches it into an up quark and in doing so transforms itself into an electron (or a muon). Such an electron or a muon emerging from a neutrino interaction immediately signals some quark alchemy. If the 1967 SLAC experiment was the analogy for quarks of the 1909 discovery of the atomic nucleus by Geiger, Marsden and Rutherford, then the CERN Gargamelle experiments were the equivalent of Blackett's 1925 observation of an induced nuclear transformation.

By 1972, using a few thousand photographs (a mere handful in bubble chamber terms, and much more economical in terms of particle interactions than the SLAC electron experiment) the new Gargamelle proton picture confirmed that the charge assignments of the nucleon partons were the fractional ones suggested by Gell-Mann. The partons in the nucleon box began to look slightly more like quarks. One problem with the SLAC electron data had been in trying to count how many partons there were in each nucleon 'box'. According to Gell-Mann's picture, the nucleon contains just three 'valence' quarks which give it its basic properties. However the electron beam saw more than just three partons, as transient quarks and antiquarks continually buzzed around inside the nucleon. Although these additional quarks and antiquarks come and go rapidly, at any one time several of them are there and an incoming electron or neutrino finds it hard to distinguish them from the permanent valence quarks. However, a synthetic neutrino beam is in fact a mixture of neutrinos and their antiparticles, namely antineutrinos. Looking at an interaction photograph, physicists can tell whether it was caused by a neutrino or an antineutrino. The neutrino interaction rate seen in Gargamelle was about three times that of antineutrinos. In 1969, theorists David Gross (from the USA) and Christopher Llewellyn Smith (from the UK), working at CERN, showed how subtracting the antineutrino effects from the neutrino effects would eliminate all the transient stuff and isolate the signal due to the valence quarks. Applying these ideas to the initial 1972 data from Gargamelle showed that there were just three valence quarks. The identification of partons with quarks was complete, and the old picture of the nucleus as a cluster of protons and neutrons held together by pions was wide of the mark, except for limited use at low energies.

However, there was still a paradox: when stung by high-energy electrons or neutrinos, quarks behaved like free particles, but all attempts to isolate quarks were unsuccessful. How could quarks behave as though they were free but still be inaccessible?

At the time, the Gargamelle picture of the nucleon was the most important new physics result to have emerged from CERN, the laboratory's first major physics discovery. There had been significant results before (precision measurements, new resonances, etc) but the Gargamelle neutrino x-ray, together with its theoretical interpretation, finally reconciled Gell-Mann's static picture of the nucleon as a composite of three quarks with the dynamic experiments which had seen nucleon substructure. This discovery alone would have been a triumph for Lagarrigue, who unfortunately died in 1975, but Gargamelle had an even more important role yet to play.

GLUED FROM THE SPOT

In this chapter, a few sections are printed in smaller type. Although these explanations are essential to an understanding of contemporary quark field theory, they are more difficult. Some readers might prefer to omit them and to pick up the story again where the text reverts to its normal size.

Electromagnetic forces can either push or pull; dissimilar pairs of electric charges or magnetic poles are attracted to each other, while pairs of charges or poles repel. This push or pull is moreover inversely proportional to the square of the distance between the charges or poles; halve the distance between them and the push or pull is four times as great. Electric charge cannot come in arbitrary quantities; it is ultimately 'quantized', i.e. divided into discrete units each carrying a basic unit of charge, the charge of the electron. Quarks have still smaller electric charges, but the charges are still quantized.

In another familiar force of nature, gravitation, the attraction between two objects is proportional to their masses, but again inversely proportional to the square of the distance between them. In gravitation, mass plays a role analogous to electric charge. Unlike electricity and magnetism, however, gravitation is always attractive; it is always a pull and never a push. As far as we know, every piece of matter in the Universe attracts every other piece.

What is it about electromagnetism that makes it both push and pull? Electricity has the idea of electric charge, which can be either positive or negative. Left to itself, an electrical system tries to become more stable, rolling down an energy slope until it reaches the bottom, where there is a minimum of electric charge. For a system of two equal and opposite charges, the bottom of the slope is when the charges come together and cancel out. Bringing together two like charges is hard work, like running uphill. Left to themselves after having been initially brought together, the like charges will move away from each other, diminishing the net local charge. Electricity tries to avoid concentrations of charge. In gravitation, there is no such thing as negative mass; so gravitation is a unidirectional attraction. Matter likes to come together.

Electromagnetic attraction and repulsion is vividly displayed in lines of force, the patterns taken up by iron filings when spread on a sheet of paper held above a magnetic field. For the attraction between two unlike poles, the lines of force run from one pole to the other; the stronger the attraction, the denser are the lines. With like poles, the lines of force veer away from each other and the density reflects the force of the repulsion. The stronger the push, the fewer lines

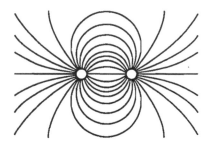

Figure 29: *Magnetic lines of force in field: for the attraction between two unlike poles, the lines of force run from one pole to the other: the stronger the attraction or repulsion the denser are the lines. These force lines map the 'field'—the sphere of influence surrounding the poles.*

of force there are at the centre of the repulsion. These lines of force map the 'field'—the sphere of influence surrounding the poles, or charges.

In quantum terms, the field revealed by these iron filings is carried by an invisible swarm of 'virtual' photons buzzing round the source of the charge. As free particles, photons (packets of electromagnetic energy in the form of light rays or their analogues (x-rays, gamma rays, etc)) can travel huge distances, but the virtual photons buzzing round electric charges or magnetic poles are not free to roam at will. These photons are instead captive, created from 'borrowed' energy which has to be returned. Each virtual photon has a quantum-mechanical mortgage which governs how long and how far it can roam around its parent charge before it has to repay its energy debt. These quantum-mechanical mortgages carry no interest. Simply, the larger the loan, the faster it has to be repaid. The energy loan multiplied by the repayment time cannot exceed a fixed amount, Planck's constant.

However, when two charges approach, their respective swarms of virtual photons start to overlap. The virtual photons now have an alternative destination and can 'forget' their allegiance to their parent charge. Rather than returning to base, virtual photons can repay their energy debt elsewhere. When this happens, the energy borrowed from one charge is transferred to another. This energy transaction between the two charges is felt as a force.

The detailed physics bookkeeping of these transactions is handled by the mathematical accounting system of 'field theory'. The quantum behaviour of the electron was enshrined by Paul Dirac in 1928. After this monumental work, Dirac set out to develop a full quantum picture of electrodynamics, but the mission only became complete with the post-war work of Feynman,

Schwinger and Tomonaga, which finally provided a framework to keep track of the charged particles and the energy debts of their surrounding virtual photons. Electrodynamics depends only on differences of electric and magnetic 'voltages', and not their absolute values. The equations of electromagnetism can therefore be cast in many ways, but the outcome is always the same: the mathematical picture is said to be invariant. Whenever there is such an invariance, the accounting system of the field theory reveals energy transfers—forces. Forces are Nature's way of preserving such underlying invariances.

> Physicists call these pictures 'gauge theories'. The word gauge stems from a brave attempt by the German mathematician Hermann Weyl in which he tried to combine together gravity, as described by Einstein's then new general theory of relativity, and electromagnetism in a single unified theory. General relativity had revolutionized physics thinking, and Weyl's idea was that changes in electromagnetic energies should correspond to a change in a standard length, or gauge, which in general relativity might vary from one point to another. While Weyl's attempt failed (general relativity and electromagnetism cannot be linked in this way), the name 'gauge theory' nevertheless stuck and is now applied to all theories which in principle can be constructed in arbitrary ways, but where the result is nevertheless invariant, independent of the arbitrariness. A rotation in three-dimensional space, for example, is not a gauge transformation, because it is specified by three definite parameters: the angles of rotation about each axis. However, in electrodynamics, the electric potential difference (voltage) between two points depends only on the position of the points and not on the path taken between them. It is the invariance of the equations under this arbitrariness which gives rise to electromagnetic forces.

At the quantum level, the interaction between two passing electric charges is not a once-and-for-all effect. Instead it is the net result of many different energy transfers. Like a game of football or a boxing match, the eventual outcome follows a myriad of individual moves. There is no quantum 'scoreboard'; the outcome, saying who wins or by what margin, can only come after examining the complete 'action replay'. Reflecting the workings of the field itself, the mathematics of quantum field theory provides a faithful step-by-step replay of an encounter. In this reconstruction, a vital parameter is the strength of the 'punch' carried by a single virtual particle. In field theory, this translates as a 'coupling', and for electromagnetism is a small number, 0.0073 (usually expressed as 1/137, dating from the 1920s when some physicists and cosmologists hoped the reciprocal of an integer number might have some deep significance).

Quantum mechanics pioneer Wolfgang Pauli was intrigued by this number and repeatedly stressed that its true meaning would lead to new understanding. The irascible Pauli could be intolerant and his criticism could be merciless, earning himself the unofficial title 'the scourge of God'. When Pauli died in 1958, his colleagues imagined Pauli's entry into heaven, with the Deity welcoming him at the gates:

'Professor Pauli, now that you are with us, do you have any specific wish to make your stay more agreeable?'

'Yes,' replied the physicist, 'I would like to know the significance of the electromagnetic coupling 1/137.'

'Very well,' replied the Deity, summoning his physics angel, who, with a pen of gold, wrote down a few equations.

'Ganz falsch' ['Utterly wrong'], thundered Pauli, storming off.

Having such a small number opens up special ways of doing calculations which can short-circuit the need to have to run through the entire field-theoretic action replay. Imagine the infinite series $1 + x + x^2 + x^3 + \ldots$. If $x = 0.1$, the series becomes $1 + 0.1 + 0.01 + \ldots$, and sums to $1.111\ldots$. The series is quickly convergent, and even taking the first term provides a useful approximation, with subsequent terms providing the successive 'perturbations' to improve the calculation. If $x = 0.0073$, as in electrodynamics, the convergence is even better, and even the first term is fairly precise—a mathematical 'knock-out punch'. In this way, quantum electrodynamics can be very selective about the mechanisms that it takes into account. With such a small coupling, this technique, known as 'perturbation theory', allows a blow-by-blow account of the interactions between two electrons to be edited to include only a few exchanges. One knows how the 'goals' will be scored, and the remaining insignificant interactions can be safely discarded. With a few relatively simple calculations, pioneers were able to account for the tiny discrepancies in the hydrogen spectrum and in the electron's magnetism.

What of other forces and their fields? What binds the quarks together to form subnuclear particles? An up quark carries two-thirds of a unit of positive electric charge, while a down quark carries a third of a unit of negative electric charge. A proton is made up of two 'ups' (which repel each other electrically) and a 'down', while a neutron has one 'up' and two electrically repulsive 'downs'. Electrical forces increase as the square of the distance separating them decreases. The electrical forces between neighbouring quarks are immense— almost a billion billion times more powerful than the force that binds electrons to atomic nuclei. Nevertheless the forces that bind together three quarks dwarf even such mighty electrical repulsions. At the quark level, gigantic new forces have to come into play. In the 1950s, physicists started to look at possible field theories of subnuclear forces, with subnuclear particles surrounded by clouds of virtual mesons. Before the appearance of quarks on the scene, any such attempts to understand the workings of subnuclear mechanisms were futile.

Looking at Gell-Mann's original quark assignments, in 1964 O W ('Wally')

Greenberg proposed that quarks, as well as electric charge, strangeness, etc, also carry another charge label which, rather than existing in two diametrically opposite kinds, as in electrically positive and negative, comes instead in three different varieties. Greenberg's original idea was to help the quark picture to sidestep some problems. Some of the initial subnuclear particle quark assignments, notably the omega minus, with its three strange quarks, violated basic rules about how many similar particles could share the same quantum accommodation. According to these rules, the omega minus, which had dramatically established Gell-Mann's quark picture, should not exist. Clearly something else was happening to enable the omega minus to overcome these objections. As time went on, the additional threefold quark label was seen to offer more and more advantages. In 1971, Gell-Mann, working with Harald Fritzsch, a young East German physicist who had made a daring escape to the West, showed how the additional quark label played other fundamental roles.

Figure 30: *In July 1968, East German physics student Harald Fritzsch made a daring escape to the West. His 1971 work with Murray Gell-Mann showed the importance of 'coloured' quarks. Fritzsch has subsequently written several science books and in 1995 received the German Physical Society's medal for science writing.*

In July 1968, physics student Fritzsch had announced he was leaving Leipzig, East Germany, for a Bulgarian Black Sea camping and boating holiday. From the coastal resort of Varna, he embarked in his fragile craft, specially equipped with a tiny outboard motor, sailing far out to sea to avoid the Bulgarian coastal watch. After a southward journey of some 200 km, exhausted and with his flimsy boat awash, he landed at Igneada, Turkey, the first time anyone had successfully escaped from behind the Iron Curtain via this route. As well as planning a new life in the West, Fritzsch had also nurtured physics ideas, including applying field theory ideas to quarks. These plans, too, were to come to fruition, but only after Gell-Mann's Pasadena office had been rocked by an earthquake just before Fritzsch's arrival in January 1971.

Handling three different quantities which mutually cancel requires special mathematics, but an appealing idea was to compare the three possible quark labels to the primary colours red, green and blue, which when they come together produce 'colourless' white light. The quarks are of course not actually coloured, but the analogy makes the idea easier to visualize. Colour is like a generalized form of electricity. As with electric charge, where opposites attract, opposite colours, like a red quark and an 'anti-red' antiquark, also stick together. Like colours, such as two red quarks, repel, but so do certain dissimilar colours. Other dissimilar colour combinations attract. As well as depending on colour, the force between two quarks also depends on the way that the quarks are put together.

The strongest attractions, or pulls, providing the most stable systems, occur when the individual colour quark charges add together to give a colourless result. This is only possible with three quarks, as in protons, neutrons and their heavier cousins, or in quark–antiquark pairs, where colour and its opposite anticolour mutually cancel, as in pi and other mesons. All other quark possibilities give a net coloured result and are less stable. This intrinsic triplicity of quark charge explains why quarks form composite triplets. Colour also nicely complements the idea that quarks have fractional electric charge. Although the constituent quarks have both colour and fractional charge, the particles that they make up carry no net colour, while the net electric charge is an integer number. This intimate relationship between colour, electric charge and other charge-like quantities is surely a key to a deeper understanding of physics.

The particles buzzing around quarks and carrying the influence of the inter-quark force, or field, make the quarks stick together. As befits a vocabulary which already displays vivid imagination, these sticky particles were called by Gell-Mann 'gluons'. Gluons, like photons, are massless but do not 'feel' electric charge. Because gluons carry messages between quarks of different colours, each gluon has two colour labels, making nine possible gluon–colour combinations. However, one combination turns out to be colourless, leaving eight kinds of bicoloured gluons, each carrying a different colour message. The fact that the gluons themselves carry charge-like labels makes the colour force look very different from electromagnetism.

With each gluon carrying two colours, a totally new possibility opens up: three different colour-coded gluons meeting at a point. This ability of gluons to interlink is in stark contrast with electromagnetic lines of force, which can never

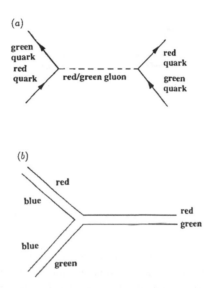

Figure 31: *(a) Inter-quark communication via a gluon. The gluon messenger has two colour 'hooks', each of which links to a separate quark leg, so that colour is transferred from one quark to another. (b) Each gluon messenger of the inter-quark force carries two colour labels. In this way, three different colour-coded gluons can meet.*

cross. When two unlike electric charges are separated, the communicating lines of force become less dense, and the pull between the charges becomes correspondingly weaker. In the quark case, if two neighbouring quarks are moved apart, the gluon paths which join them are separated, but this opens up the possibility that additional gluon paths jump in to join the original gluon paths. The further apart the quarks are moved, the more and more gluon lines are added, until the inter-quark space becomes filled with a dense net of interconnected gluons. Thus, with quarks, the farther apart they are, the stronger is the force between them. Conversely, as the quarks get closer together, fewer and fewer gluon connections are possible, and the force becomes weaker. However, when quarks are pushed apart, after a certain stage the intervening gluons become so opaque that the quark landscape is totally obscured by gluons. The quarks can no longer 'see' each other, and the inter-quark force reaches a limiting value.

However, this gluon screen can be sidestepped by taking quarks 'by surprise'. If a high-energy electron or neutrino slices through the surrounding subnuclear

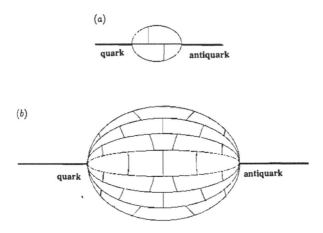

Figure 32: *Initially (a), this neighbouring quark and antiquark are linked by just a few gluon lines of force. However, as the quarks move apart, the interconnecting gluon net becomes progressively more intricate, adding to the force binding them together (b). Paradoxically, the more effort expended in separating the quarks, the stronger the links become. It is futile to continue.*

husk, the quark is caught unawares. High energy intruder particles can see through the otherwise opaque nuclear glue. Taken by surprise, the quark recoils from the intruding electron or neutrino before the gluon bonds with the neighbouring quarks have a chance to rearrange themselves and become strong. Conversely, if the attack is less stealthy and the inter-quark bonds are severely rattled before the incoming particle even reaches the quark, the gluon links have time to prepare themselves and to protect their quarks even more securely, throwing up a intervening self-defence mechanism which is difficult to penetrate. A quark can be sniped at but is impervious to heavy bombardment. In 1967, SLAC's high-energy electrons succeeded in hitting quarks by sniper fire, qualitatively explaining why the quarks seen in the 1967 SLAC experiments behaved like free 'partons' rattling around freely inside a box. The challenge was to formulate a mathematical theory of the colour field, analogous to quantum electrodynamics, which made accurate calculations possible.

In quantum electrodynamics, the result of two underlying invariance operations is independent of the order in which they are applied; the theory is said to be 'Abelian', and this dictates our everyday experience of electromagnetism. An example is given by reflection by ordinary plane mirrors. Reflection in a single mirror gives a left–right inverted image, but reflection in a second mirror restores the original image, irrespective of which mirror is used first. However, the more general case is when the theories are 'non-Abelian', when the result of applying two consecutive operations depends on the order they are performed. An example is two consecutive rotations; the final orientation of a coin which is flipped and then slightly rotated is different to that of a coin which is first slightly rotated and then flipped.

In field theory, the importance of tractability or 'renormalizability' had long been realized. Physicists could write down what appear to be compelling field theories but, when they tried to use them, troublesome infinities clogged up the calculations. This had bothered quantum electrodynamics throughout the 1930s, and Dirac had given up in despair. In the blow-by-blow account of an interaction, these infinities occurred when the electron 'hit itself'—the interaction of the electron with its own virtual photons. It was as though a football player inadvertently putting the ball in his own goal gave the opposing team an infinite score, and there was no point in continuing the game. However, once Feynman, Schwinger and Tomonaga had shown how quantum electrodynamics could be renormalized (systematically purged of such troublesome log-jam infinities) the results began to flow.

This recipe for renormalization only worked because of the smallness of the electromagnetic coupling, 0.0073. For field theories with more powerful couplings, it looked as though the infinities came back, rendering predictions impossible. Even if the infinities could somehow be reconciled by a new recipe, the individual quark interactions looked to be extremely strong, and the field theory blow-by-blow account could not be edited down to a few exchanges. In quark play, the goals would occur too frequently to count. Solving a simple quark interaction looked like a commitment to a lifetime of computation. For a time, just as in the 1930s Dirac had despaired of a tractable formulation of quantum electrodynamics, many physicists despaired of quark field theory.

However, in the early 1950s, physicists had formulated a more general approach to renormalization. Although this did not immediately open the way to calculations in any particular field theory, it enabled different field theory scenarios to be explored and compared. A basic problem in any field theory is the interaction of a single particle with itself, its own field. A lone electric charge is surrounded by its own virtual photons and a lone quark is surrounded by gluons. These surrounding clouds mask the particles seen from afar; so they do not appear as hypothetical 'bare' particles, stripped of their attendant cloud of photons or gluons. In fields with powerful couplings, such as those of gluons, these masking effects make the apparent coupling strength under one set of conditions look very different from that under other conditions. The general renormalization techniques can trace the evolution of these couplings.

The quark idea had been in vogue following the discovery of the omega minus in 1964 but, with predictions coming via other routes, quark field theory had not been pushed hard. However, in the early 1970s, with the existence of quarks inside protons looking more likely, and the discovery by Gerard 't Hooft at Utrecht that non-Abelian field theories were indeed renormalizable, physicists returned to the idea of a field theory of subnuclear forces. Perhaps a field theory of quarks was within reach.

In the early 1970s, theorists in Europe and the USA blazed parallel trails through the field theory jungle, an effort which culminated in 1973 when David Gross and Frank Wilczek at Princeton and David Politzer at Harvard saw that in certain (non-Abelian) theories the behaviour of the contributions from gluon-like carriers dictated that the effective strength of the field decreased at short distances. The closer together the particles, the weaker is the force between them. Ultimately the influence of the force disappears completely! Conversely, as the particles become farther apart, the force gets stronger. This bizarre 'asymptotic freedom' contradicts commonsense experience but provides a natural framework for a picture of quarks interacting through coloured gluons. It means that the gluon, although it is massless and at first sight could roam as far as the photon of electromagnetism, is condemned by the colour force to operate only at short range.

With the end result unknown, these pioneer calculations were not straightforward and algebraic errors hindered the outcome. However, today they are given as homework problems—the theoretical analogy that yesterday's experimental discovery becomes tomorrow's troublesome background.

With the discovery of asymptotic freedom, the way was open for a full field theory of quarks and gluons. Embedding colour in such a mathematical field theory of quarks and gluons gives quantum chromodynamics, after the Greek word 'chromos', for colour. The name, usually abbreviated to QCD, is Murray Gell-Mann's suggestion, intended to stress the analogies with quantum electrodynamics, together with the colour-coded effects of the carrier gluons. The strength of the gluon effects means that the quark coupling is not a fixed number, varying instead according to the prevailing conditions. Experiments show that the coupling of quark interactions measured on the scale of laboratory experiments does indeed vary but seems to converge to about 0.1, much larger than the 1/137 of electromagnetism, but still small enough for a blow-by-blow reconstruction of a quark encounter to be edited to a few frames.

Comparing two sets of measurements must be handled very carefully but, if this is done correctly, the variations are themselves evidence for quark–gluon effects. The SLAC experiments with high-energy electrons had revealed the proton as a box of partons, subsequently identified as quarks, using the quark-sensitive neutrino beams. However, the quarks-in-a-box picture is only an approximation, and comparing results under different conditions reveals slight variations from the simple box picture, which can be accounted for by QCD calculations. In these interactions, more important than the energy of the incident electron or neutrino is the energy transferred from the incident particle in an interaction with a quark. Tracking this energy transfer, physicists were clearly able to see how this shifts the distribution of quarks, an effect predicted by QCD. For the first time, physicists were able to quantify quark dynamics. Using neutrino beams, physicists were also able to compare the behaviour of up and down quarks inside protons and neutrons. Again the slight difference is explained by delicate colour 'magnetic' effects. Just as a spinning electric charge behaves as a tiny magnet, so a spinning colour charge produces additional forces. As well as explaining the different properties of up and down quarks inside subnuclear particles, this colour magnetism makes different quark assignments have different masses. Subnuclear particles where the three quark spins are aligned in the same direction, as in the decuplet containing the famous omega minus, are heavier than counterpart particles in which the quark spins are aligned in different directions.

With the interior quark spectrum of protons completely mapped, the final task was to add up the momentum contribution carried by the individual quarks. The result came to about 50%; only about half the momentum of a proton or neutron is carried by its component quarks. Although it is the quarks which completely determine the nature of the subnuclear particle, its electric charge and other quantum numbers, the particle contains much more besides. This is the gluon contribution. Carrying no electric charge and completely transparent to neutrinos, gluons are invisible to electron and neutrino beams. Gluon effects show up when quarks collide with each other, conditions first realized at CERN's ISR in the early 1970s. However revealing the gluon was to require experimental resourcefulness.

With the quark picture of nuclear constituents and the formulation of QCD (the field theory of quarks and gluons), the understanding of subnuclear forces became complete. Yukawa, with his proposal for a heavy carrier of the strong force, had been on the wrong track. The forces binding subnuclear protons and neutrons together in nuclei are now understood as being analogous to the interlocking quantum bonds which hold electrically neutral atoms together in complex molecules. Atomic nuclei, once considered as the bedrock of matter, are in fact complex quark molecules. Just as the dynamics of complex molecules is difficult to understand in terms of individual atoms, so the behaviour of complex nuclei is far removed from quark physics. At every scale, Nature displays an initially unsuspected richness and complexity.

Inventing the gluon and explaining how a massless particle could explain the short-range force which bound quarks together required considerable mathematical and theoretical ingenuity. The field theory avenues which had been explored in this quest also opened the door to another puzzle, explaining how one type of quark could transform into another, and how such unfamiliar transformations subtly complemented the familiar world of electromagnetism. Just as asymptotic freedom and the apparently paradoxical behaviour of quarks had called for new breakthroughs in understanding, so this other aspect of quark behaviour was to show that Nature had another trick up its sleeve.

IDEAS IN COLLISION

In April 1960, an airplane carrying CERN Director General Cornelis Bakker crashed on its approach to New York's Idlewild Airport. For the second time in its short history, CERN unexpectedly found itself without a leader. Edoardo Amaldi made plans to leave Rome to take charge of CERN affairs temporarily, but on arriving at the Geneva laboratory found that John Adams, who five years previously had stepped into the shoes of Frank Goward as head of the proton synchrotron group, had this time stepped in as emergency Director General. Everything seemed to be running smoothly, and a special meeting of CERN Council in May confirmed John Adams as Acting Director General, to remain in office until August 1961, by which time a suitable replacement should have been found. However, Adams' talents were also in demand elsewhere; he had also been asked to head the new UK thermonuclear fusion laboratory at Culham, near Harwell.

Despite commissioning the world's highest energy proton synchrotron in November 1959 and beating Brookhaven to the post, CERN did not capitalize on this initial success. Too much emphasis had been placed on bringing the new machine into action, while little attention had been paid to how it should be subsequently used for physics. CERN had somehow expected that university teams would come beating on its door, but they did not. European physicists were not used to having a big synchrotron on their doorstep and little had been done to attract them. On the other side of the Atlantic the situation was very different. The early 1960s were to see Brookhaven's new machine make several major physics discoveries in succession; the pioneer neutrino experiment by Lederman, Steinberger and Schwartz in 1962 (Nobel prize 1988), Samios' omega minus in 1963, and an experiment by Val Fitch and Jim Cronin in 1964 which showed an unexplained effect in the symmetry properties of neutral kaons and suggested that, if the arrow of time were reversed, physics would look different (Nobel prize 1980). CERN, with a similar synchrotron, most of the time could only sit and watch as the Americans made one major discovery after another. However, during his brief spell as Acting Director General, John Adams acknowledged these deficiencies at CERN and established a new committee structure to invite proposals for new experiments from the physics community and to select and develop the best among them. This flexible system is still in use and has produced historic experiments but in the early 1960s needed time to bear fruit.

CERN's first problem was to find a new Director General. One outstanding candidate was Victor Weisskopf. Born in Vienna in 1908, Weisskopf had witnessed and collaborated in the development of quantum mechanics in the 1920s as a graduate student in Max Born's school in Göttingen, where Ehrenfest, Heisenberg, Jordan, Pauli and Wigner, among others under Born's guidance, had carved out the new theory. After Göttingen, Weisskopf had worked with Schrödinger in Berlin, with Bohr at Copenhagen, and then with Dirac and Peierls at Cambridge, before being invited by Pauli to be his assistant in Zurich. In 1937, Weisskopf moved to the USA, working at Rochester with Hans Bethe. During these years, Weisskopf greatly influenced the development

Figure 33: *Victor Weisskopf, CERN Director General from 1961 to 1965, transformed the new laboratory from a venture in international collaboration into a productive research centre, instilling new spirit and setting it on a firm course towards success.*

of quantum field theory, but the Second World War saw an abrupt change in career path. When Bethe was appointed Leader of the Theory Division at Los Alamos, Weisskopf was made Bethe's deputy. Los Alamos gave Weisskopf plenty of opportunity for seeing scientist–managers such as Oppenheimer in action and this experience was to influence Weisskopf's style of management greatly. However, after the war, he moved to MIT, returning to unfinished business in quantum field theory. In 1947 he was one of the discussion leaders at the historic conference on Shelter Island, New York, from which the modern theory of quantum electrodynamics first emerged. In the 1950s, Weisskopf began to look back to Europe and was increasingly seen on that side of the Atlantic. He had been invited by Bakker to join the CERN Directorate in 1958 but had refused. 'We had different ideas about how CERN should be run,' he wrote later.

Many people were convinced Weisskopf was the right man for CERN. He was European by birth, he was a distinguished physicist, and he was highly motivated. In December 1960, John Adams, on behalf of CERN, offered Weisskopf the post of Director General of the European laboratory. Initially Weisskopf wavered at accepting such heavy responsibility but finally accepted. As leader of the laboratory, he realized that he would be in a position to implement his own judgment, to do what he thought was right, rather than to execute what others thought was right. This decision, like many of Weisskopf's subsequent judgments, turned out to be right. Weisskopf transformed CERN from a venture in international collaboration which had built a remarkable new machine into a productive scientific laboratory whose lifeblood was ingenious innovation, inspired research and far-sighted planning. Weisskopf instilled spirit into CERN and set it on a firm course towards success.

Weisskopf's initial aim was to spend some time at CERN to learn how it worked before gently moving sideways into the Director General's seat. Unfortunately soon after his arrival in Geneva early in 1961 he was injured in a car accident and spent a long time in hospital, eventually having to return to the USA for specialist surgery, which prevented him from learning as much about CERN as he would have liked. With John Adams departed for Culham, Weisskopf chose as his closest collaborator Mervyn Hine, who had played a vital role with Adams in building and commissioning CERN's new proton synchrotron.

One area where the CERN proton synchrotron had tried to make a breakthrough and initially failed was in neutrino physics. Under Weisskopf, remedial action was taken. New beam extraction techniques were developed, Simon van der Meer invented his magnetic horn, and plans began to develop for suitable detectors, including Lagarrigue's big heavy-liquid bubble chamber. Weisskopf realized the potential in this distinctive line of research and took neutrino physics at CERN under his wing. The laboratory had lost the first round to Brookhaven, but there would be many rounds yet to come.

The new proton synchrotrons at CERN and Brookhaven had broken new ground in accelerator design and signposted the beginning of a new road. Where could this new accelerator road lead? In the USA, accelerator progress had helter-skeltered from the cyclotron to the strong focusing synchrotron in just fifteen years, with the synchrocyclotron and the weak-focusing synchrotron milestones along the route. If this heady rate of progress was to continue, who could dream what the next fifteen years had in store? Accelerator development had progressed leapfrog style, with new projects taking off over the back of existing ones. While in the USA these new accelerators were scattered all over the country, in Europe CERN had become the natural focus. CERN's twin initial objectives had been for a 600 MeV synchrocyclotron to be built by Bakker's group while the much bigger proton synchrotron took shape. However even before the 600 MeV machine was complete in 1957, attention began to turn towards longer-term goals. In 1957, an accelerator research group was set up by John Adams with the task of exploring new machine ideas. This small group was reinforced when the proton synchrotron began operation in 1959 and machine specialists, no longer preoccupied with operational problems, could look to the future.

Accelerator specialists were facing new challenges. With the new strong-focusing machines having raised the energy horizon, attention had turned to adapting synchrotron designs to provide more particles. As understanding advanced, more and more physics was charted in detail. Discoveries, by their very nature, are rare; many thousands of bubble chamber photographs had to be amassed before new effects were found. Rarity could be compromised by pushing the accelerator, pumping out more particles per second. If accelerator output could be increased a hundredfold, what would otherwise take a year to discover could be compressed into a few days of running.

Another approach was to collide particle beams together. When the particles from a synchrotron slam into a stationary target, most of the precious projectile energy is lost in the target recoil, and only a fraction actually goes into the collision. If instead two beams could be fired at each other, all the kinetic energy would go into the collision, providing a considerable increase in collision energy and physics potential. In 1956, Gerry O'Neill of Princeton proposed a way of combining these two goals of intensity and colliding beams, using an ordinary synchrotron to accelerate the particles, and then accumulating them in two rings which met tangentially, so that the two stored beams could be brought into collision.

The Midwestern Universities Research Association (MURA) consortium, based in Madison, Wisconsin, was also looking into these ideas for colliding proton beams. O'Neill, meanwhile, was proposing a similar scheme for electrons, which, being much lighter than protons, are much easier to accelerate. An electron–electron collider was built at Stanford by a Stanford–Princeton collaboration. Another electron–electron machine was built at the Novosibirsk laboratory recently set up by the mercurial Russian physicist Andrei Mikhailovitch Budker. Born Gersh Itskovitch of Jewish extraction, Budker, who

Figure 34: *The pioneer VEP-1 electron–electron collider, built by the mercurial Russian physicist Andrei Mikhailovitch Budker, came into operation at Novosibirsk in 1965.*

was married five times ('all my romances ended in marriage'), changed his name because he wanted a title which sounded 'more Russian'. All his life Budker pushed for the development and application of particle beams, and his Novosibirsk laboratory became a temple dedicated to the furtherance of particle beams, with Budker as its high priest.

Meanwhile the Austrian Bruno Touschek, working at the Italian Frascati laboratory, had another idea which was to change the face of colliding beam machines. Touschek, like Budker, died before reaching the age of 60, but both men left machine physics monuments to their own memory. Positrons have the same mass but opposite charge as their antimatter counterparts, electrons. Thus in the same magnetic field, electrons and positrons of the same energy circle round at the same rate and in the same orbit, but in mutually opposite directions. Instead of colliding electron beams held in two distinct rings, suggested Touschek, why not make electrons and positrons go round the same ring in opposite directions? The circulating electron and positron beams could be kept slightly apart but nudged together to collide at designated points. As well as using a single storage ring instead of two and saving on construction costs, this approach had additional physics attractions. While electrons just electromagnetically scatter off electrons, electrons and positrons mutually annihilate and can form particles composed of quarks and antiquarks bound together. A machine colliding subatomic electrons and positrons can be used to study subnuclear particles. The first electron–

positron storage ring, Annello d'Accumulazione (AdA), with a circumference of 4 m, stored its first beams at Frascati in 1961 and was subsequently used for electron–positron collisions at the French laboratory at Orsay, near Paris. The electron–positron idea caught on fast, and new projects were developed at Cambridge (Massachussetts), Stanford, Novosibirsk, Orsay and Frascati. The new electron–positron collision path through the physics jungle became a major route.

These compact pioneer electron–positron colliders were in sharp contrast with the big proton machines. The advent of large proton synchrotrons had meant that physics needed substantial purpose-built laboratories which were difficult for individual universities, and even nations, to construct. These big new synchrotrons, several hundred metres in diameter, were impressive by their sheer size. Machines with futuristic names such as 'Cosmotron' naturally attracted media interest, and it is a pity that many subsequent, and much bigger, machines have been handicapped by names which lack flair and imagination.

Although large, the new proton synchrotrons were still small enough to walk round. In June 1959, delegates at an accelerator conference at Madison, Wisconsin, were startled to hear a proposal from Matthew Sands of Caltech for a synchrotron several kilometres in diameter to attain the unheard of energy of 300 GeV. To minimize the cost of its major component, the magnet, the giant synchrotron would have to be fed by a smaller machine of more conventional proportions. As yet, few took Sands seriously; the new CERN and Brookhaven strong-focusing machines had not yet even been commissioned. However, the seed of a multihundred GeV machine had been sown, and the giant machine idea began to be a regular feature of international accelerator meetings. Giant machines were not entirely new; Stanford had already introduced the idea of a 2 mile linear accelerator and, where linear machines could go, why not circular ones too? By the summer of 1961, a 150-page Brookhaven report described a design study for a 300–1000 GeV accelerator. With such large super-synchrotrons on the cards, it was clear that CERN would have to give some thought to following suit if it was to maintain its recently established position at the front of the particle accelerator pack.

Meanwhile the accelerator research group at CERN had already latched onto the beam stacking idea proposed by MURA, and an electron model was built. An initial 1960 proposal foresaw two adjacent 25 GeV proton storage rings touching tangentially, but this was soon transformed into two interlaced storage rings: the Intersecting Storage Rings (ISR). This was promoted at CERN, particularly by Mervyn Hine, as a natural extension of the capabilities of the existing synchrotron, rather than a totally new project. This was because the CERN Convention, the laboratory's constitutional document, had foreseen a 'basic' programme, supported by all member states but allowed the possibility of 'supplementary programmes' with special financing.

In one of his final papers as Acting Director General in 1961, John Adams foresaw a choice: the natural extension to the ISR or a new super-machine. A new supersynchrotron would have to be launched as a major new CERN project and go the round of national participation all over again, with the risk that one or other of the CERN Member States might opt out. In addition, there were geographic implications which could easily become political crusades.With the CERN site, sandwiched between the Jura mountains and Lake Geneva, far from flat, from the outset the natural assumption was that such a supersynchrotron would have to be built elsewhere, but nevertheless within the CERN contractual umbrella. It would be another branch of CERN, CERN II, established somewhere else.

Subsequent discussions were curiously sceptical about the attractions of a super-synchrotron. While a good case could be put for extending the known energy frontiers, it was also seen as a 'brute force' extension of existing techniques which could find itself overtaken by new developments during its relatively long construction time. The situation began to resemble that of ten years before when the first scientific objectives for CERN had been discussed. Then, a conventional synchrocyclotron emerged as the optimal short-term goal, with the 'big machine' following later. Why not do the same again? The ISR, while certainly innovative, was restricted in the physics that it could do. It was seen by many scientists not as an accelerator, but as a big proton–proton physics experiment.

With Weisskopf installed in the CERN driving seat, this two-pronged aspiration had to be resolved. To do it democratically, and to ensure that the resulting research programme was optimally suited to its users, the physicists, rather than to the convenience of CERN's management, a European Committee for Future Accelerators (ECFA) was set up as a forum for CERN users. Under the chairmanship of Edoardo Amaldi, its first report, published in June 1963, was far reaching in its implications, recommending construction of both the ISR, as a natural extension of the capabilities of the existing synchrotron, and of a high-energy (which meant 300 GeV) machine, delivering ten times the proton intensity that the CERN proton synchrotron was achieving at the time. The users had given a clear signal that they wanted both machines. However, building a 300 GeV synchrotron at a new CERN laboratory, wherever it might be, while the existing laboratory would gear up for the ISR, proved too big a morsel for CERN to swallow in one gulp. To get accepted, the two projects had to be decoupled and, after intense debate and negotiation which shook CERN to the core, the ISR proposal was formally approved in December 1965.

While the administrative approval procedures were being hammered out, the design of the ISR had been finalized by a special study group under Kjell Johnsen. The ISR had to be built alongside the existing synchrotron which would supply the ISR particles but, with this synchrotron already right at the edge of Swiss territory, the ISR had to be built in a natural extension of the CERN site in France. CERN thus became international on the map as well as in its constitution, a development which would go on to make considerable impact.

Although the ISR had been presented as CERN's 'safe bet' compared with a 300 GeV machine, it was not entirely smooth running. The designers knew from the outset that high-vacuum technology would have to be pushed to the limit. Any scattering of the stacked proton beam due to molecules of impurity gas would spoil performance. Vacuum would have to attain 10^{-11} Torr, equivalent to the emptiness of outer space, rather than the 10^{-6} Torr obtained using turbomolecular pumps in conventional synchrotrons. For the ISR, new technology (sputter ion pumps and sublimation pumps) took over where the turbomolecular pumps left off. To transfer protons from the synchrotron to the ISR's stacking orbit called for ingenious radio-frequency gymnastics. Handling particles· in unprecedentedly close proximity produced unexpected beam idiosyncrasies which first had to be understood and then allowed for. With such delicate effects coming into play, diagnostics and control systems required special attention. It was a challenge to machine builders, but the skills that the CERN accelerator specialists acquired from the ISR were to stand them in good stead later.

In the 1960s, the US physics community also had a 'wish list' of new proposals, each of which had its own proponents. In 1962 the President's Scientific Advisory Committee and the General Advisory Committee of the Atomic Energy Commission (AEC) appointed a panel under Norman Ramsey to distil from these uncoordinated proposals the nation's future needs in high-energy accelerator physics. In May 1963, the panel ambitiously recommended (1) 'prompt' construction of a 200 GeV accelerator at Berkeley, (2) construction of colliding-beams storage rings at Brookhaven, (3) design studies at Brookhaven for a proton machine in the 600–1000 GeV range, (4) construction of a high-intensity 12.5 GeV proton accelerator by MURA 'without permitting this to delay the steps towards higher energy' and (5) development and construction of electron–positron colliding beams storage rings at Stanford—an impressive list.

The MURA proton machine proposal was the first to fall by the wayside, with no presidential support, while Berkeley soon discovered that their proposed machine was of national interest. The AEC stepped in and invited all states interested to submit site proposals for the new big machine, with a total of 125 eventually being received. By March 1966, these had been whittled down to six, in California, Colorado, Illinois, Michigan, New York and Wisconsin (where MURA had been based). To oversee preparations, a new conglomerate was set up, Universities Research Association, initially involving 34 university presidents, which eagerly awaited the outcome of the site selection process. In December 1966, the announcement came that the Illinois site, near the small town of Batavia, southwest of Chicago, had been selected. The following month,

Figure 35: *Fermilab founder Robert Wilson in 1968 with a model of a magnet for the future accelerator.*

the dynamic Robert Wilson, whose new 10 GeV electron synchrotron at Cornell had just started to operate, was invited to be director of the new National Accelerator Laboratory. Mirroring Panofsky's initial reluctance to take on responsibility for building SLAC until adequate arrangements were in place to guarantee him sufficient autonomy, Wilson, a letter of resignation in his hand, held out for special conditions which would ensure his own ability for independent action. The federal authorities blinked first and Wilson's demands were met. Early design work for the new machine was done by Wilson and his assembled consultants at Cornell before moving to temporary premises on the new Illinois site. The ambitious Wilson soon included in the design the possibility of operating at 500 GeV rather than the initial 200 GeV energy, while the completion date was pushed *forwards* from 1974 to 1972.

During this time, Weisskopf had returned to MIT after having left his indelible imprint on CERN. The European laboratory's Director General was now Bernard Gregory, a very 'British' Frenchman and a product of Leprince-

Ringuet's influential school. Gregory had long been an enthusiastic supporter of the European plan for a multihundred gigaelectronvolt machine and was happy to be in the driving seat when this machine was the immediate objective. Site offers came in from all sides, and an initial hundred or so was soon distilled down to 22 serious candidates. Asking countries to select their own major contender site reduced this to 12, with still some duplication in Germany and Italy but, by December 1967, only five possible sites remained. The CERN rules were adapted to cover two laboratories rather than one, in readiness for the new project, wherever it would be built. However, with the economic climate cooling down, only six of the 11 CERN member states were indicating their readiness to underwrite the new scheme. Undaunted, Gregory invited John Adams back to CERN to plan the construction of the new machine and its laboratory.

Reinstalled at CERN, Adams took a hard look at the machine design. The big machine being built by Robert Wilson at Batavia introduced a new magnet design, using separate dipoles to bend the beam and quadrupoles to focus it, rather than the old curved 'combined-function' magnets of previous strong-focusing synchrotrons. This increased bending power opened up the possibility of higher-energy beams for the same circumference, or alternatively a smaller circumference for the original design energy. This enabled the design team to introduce a 'missing magnet' scheme, in which a stripped-down version of the synchrotron could be installed first, operating at a lower energy, say 200 GeV. This would bring initial cost savings and, as additional money became available, the missing magnets would be added later to ramp the machine up to its full energy.

It had been generally assumed that the new machine would be built somewhere else where the ground was flat, so that the ring could first be excavated as a trench in which the housing for the machine could be built and subsequently covered over. Adams had no such preconception and looked at the possibility of a tunnel under the hilly terrain next to the existing CERN site, where an underground tunnel would have to be bored rather than excavated. Surveys of the surrounding underground terrain showed that the extent of the rock suitable for such boring could only accommodate at the most a 2.2 km diameter machine, rather than the 2.4 km originally envisaged. However, the extra cost of boring an underground tunnel could be offset against the advantages of using the existing proton synchrotron as an injector. CERN's machine had been upgraded with the addition of a 'Booster' machine to increase the proton supply, while major new experimental facilities were in the pipeline. By siting the new machine alongside the old, the new machine could share the benefit of these upgrades.

With only six member states having been in favour of the project, it was important to increase this support, particularly from Britain, which had been noticeably cool. However, in 1970 a change in government brought both a new pro-European stance and one Margaret Thatcher as Secretary for Education and Science. One of the first things that she did was to visit CERN, and the impression this made was to help to bring the needed support. The December CERN Council session brought more votes, including that of the UK, in favour of the new machines being built alongside the existing CERN site. With some nations still wavering, a final vote was not taken. When the meeting was reconvened in February, Norway, Sweden and Holland followed suit, and ten

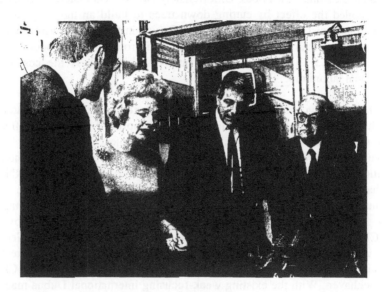

Figure 36: *In 1970 a new Conservative government in the UK brought a new pro-European stance. The visit of Secretary for Education and Science Margaret Thatcher to CERN brought UK support for the proposed big supersynchrotron to be built alongside the existing CERN site. Left to right, Peter Standley of CERN, Margaret Thatcher, CERN Director General Bernard Gregory, Edoardo Amaldi.*

nations agreed to go ahead with the new supersynchrotron at a cost of 1150 million Swiss francs. (However, in the UK the price of this new machine would also be the closure of the nation's own high-energy physics machines, and its physicists would have to rely totally on CERN.)

Gregory's mandate as CERN's Director General was drawing to its close, and Willibald Jentschke from Austria had been designated to succeed him. Although a wise decision to build the new CERN machine alongside the old one had been taken, nevertheless the administrative plan was still to have two CERNs. Jentschke became Director General of the original CERN, Laboratory I, while John Adams became Director General of the new Laboratory II, with its headquarters at Prévessin in France, 3 km away from the main CERN I site at Meyrin, Switzerland. As well as being in separate countries, the two laboratories looked very different. CERN I, with its closely spaced utilitarian concrete structures, barrack-like buildings and ugly high-rises contrasts with the greenness of Prévessin, where the buildings hug the ground, which is just as John Adams wanted it to be. One of the new beam lines was even redesigned in order to save cutting down trees. Unfortunately twenty years later, Electricité de France spoiled the effect by cutting down trees to build an ugly high-tension line. In an administrative miracle, CERN's Prévessin telephones were linked to the Swiss rather than to the French exchange, introducing such anomalies as French services with Swiss telephone numbers, but such is the cost of an international laboratory.

Meanwhile a newcomer on the world accelerator scene had appeared at Serpukhov, south of Moscow. Earlier, at Dubna, north of Moscow, a 10 GeV proton synchrotron had been built as a secret project in the early 1950s. However, it was a weak-focusing dinosaur and was soon overtaken by the new generation of strong-focusing synchrotrons, spearheaded by those at CERN and Brookhaven. It was from this Dubna machine that CERN had wrested the world high-energy crown in 1959. Dubna had also been modelled as a mirror image of CERN for Socialist countries, including those of Eastern Europe. However while CERN retained a unique international flavour although based in Geneva, Dubna was predominantly Soviet, later Russian. Eager to regain lost prestige, the Soviets pushed for a bigger machine to attain 70 GeV, outgunning CERN and Brookhaven. With the existing weak-focusing international Dubna machine unsuitable as an injector, the new Soviet machine was built instead at a new laboratory at Serpukhov, where it came into action in 1967, briefly re-establishing the Soviet position as holder of the world record for proton energy.

Across the world, at SLAC in California, the 2 mile electron linear accelerator was showing that the proton contained tiny constituents, later to be identified with quarks. While electrons can pierce to the heart of protons, clumsy proton beams could not easily see their own fine structure. Instead, the Serpukhov machine saw the overall proton envelope, the 'zone of influence' of the proton, in a new energy range. Earlier experiments had found that this zone of influence decreased as the energy increased; the proton 'shrinks' in size. If so, this had heavy implications for higher energies, where the proton might continue to shrink, becoming less and less significant as a particle. How big was the proton at the new 70 GeV horizon?

The Serpukhov machine also marked the beginning of fruitful CERN–USSR collaboration. Under a special agreement, CERN constructed special equipment for the new machine, in return for which scientists from CERN and its member states were able to participate in the new experimental programme. In the deep winter of the Cold War, such contacts were valuable in promoting East–West dialogue and overcoming political obstacles. On the scientific side, these first collaborative CERN–USSR experiments at Serpukhov showed that the proton did not shrink any more. Physicists breathed a sigh of relief. But was this new behaviour seen at Serpukhov really the beginning of a new regime for the proton, or was it merely a breathing space before the proton started to shrink again? Only higher-energy experiments would be able to tell.

The question was resolved soon after the ISR produced its first colliding beams in January 1971. Two proton beams of up to 31 GeV meeting head-on provided a collision energy equivalent to a 2000 GeV proton beam hitting a wall. With the new machine seen initially as a big proton–proton experiment, physicists were eager to see how the proton behaved at these new energies. Soon it became clear that the zone of influence of the proton was still increasing; the proton as a composite particle had not reached the end of the exploration road.

The ISR was the first major innovative particle beam facility to be built outside the USA, and some US physicists were eager to work at this new machine and experience what it was like to collide high-energy proton beams. When these beams collide, most of the collision debris is swept along with the beams; so, in the rush to get fast results, detectors were mounted as close to the beam direction as possible in initial experiments. However, in 1972, soon after the start of ISR operations, a US–CERN group chose instead to look at what came out sideways, at wide angles to the direction of the colliding protons. The group was looking for a new type of particle and figured that sideways on was a good place to look for it, as the deluge of collision debris in the beam direction would swamp anything else. Setting up its detectors away from the beams, the group and others were surprised by the unexpectedly high level of pions emerging sideways. Unprepared for such an effect and with only a modest detector, the ISR physicists were caught off-guard.

As the protons swept past each other in the rings of the ISR, most of the time it was only the fuzzy outer regions of the proton clouds that touched each other. These proton clouds continued their separate ways, with the constituent quarks deep inside the protons mere spectators. Occasionally there were real collisions; the proton clouds met head-on and the incoming constituent quarks crashed into each other. The wide-angle pions seen at the ISR came from such quark interactions deep inside the protons. The ISR quark interactions were also the first time that the quarks had been studied in their own natural habitat; the quark interactions seen at SLAC were due to the electromagnetic forces between quarks, which are electrically charged, while those interactions seen at CERN using neutrino beams were due to the weak force acting between quarks. At the ISR, the inter-quark forces came into their own and physicists were surprised how prolific the particle yields were.

While high-energy electrons and neutrinos were able to knife unimpeded to the centre of the proton, the quarks in ISR protons had to remain tightly confined so that, when ISR quarks met head-on, the quarks could not simply recoil. In these quark–quark collisions, the quark bonds inside each proton are jolted severely, and only occasionally is this jolt strong enough to break the quark bonds. These colour forces between quarks are like a bar magnet; breaking it in half does not separate the two original poles; the broken-off ends immediately become poles in their own right. Likewise, when a quark bond is broken, fresh quarks appear on the ends of the broken bonds. These emerging pieces of quark elastic are seen as characteristic 'jets' of particles, flying off in the direction in which the quarks bonds were stretched.

The big detectors needed to intercept particles emerging sideways to the beam direction, and analysis of these jets did not appear until later in the ISR programme. Had they been there at the outset, then ISR's impact on physics would have been more dramatic, but the ISR had been conceived and presented as a logical extension of CERN's proton synchrotron. Although universally acclaimed as a masterpiece of the accelerator builder's art, it suffered, at least initially, from the same drawback as the machine which fed it—too much attention had been paid to the machine and not enough into what it was going to be used for. Having made this mistake twice, CERN had learnt its lesson.

On the other side of the Atlantic, Robert Wilson's new multihundred gigaelectronvolt supersynchrotron made the running. With an authorized construction budget of US$250 million, construction had begun in 1968. Under Wilson's flamboyant leadership, it took just four years to complete the 6.4 km ring and accelerate protons to 200 GeV, four years in advance of CERN's supersynchrotron. The multitalented Wilson had a flair for architecture, and his vision and imagination were imprinted all over the new laboratory. The spectacular Main Building high-rise, modelled on a thirteenth century French cathedral, is a monument in its own right. On the technological side too, Wilson was ambitious. With an eye to the future, research and development work began on more powerful magnets for a higher-energy ring in the same 6.4 km tunnel, this time using superconducting technology to reduce power consumption. After returning US$6.5 million of construction money to the US government, the new laboratory was formally dedicated in 1974 as the Fermi National Accelerator Laboratory (routinely abbreviated to Fermilab) in honour of Enrico Fermi, who had done so much important work nearby at Chicago and who had died in 1954.

Work   for   CERN's   new   supersynchrotron,   delayed   by   the   initial

Figure 37: *The writing on the wall. Top, Intersection 8 at CERN's Intersecting Storage Rings (ISR) as it was in 1970, devoid of instrumentation for experiments. By 1983, below, the scenery at ISR Intersection 8 had changed and the beam crossing point was almost surrounded by high-technology equipment to intercept the results of the ISR beam collisions.*

ISR–supersynchrotron dilemma, was completed in 1976 and had its first taste of protons on 3 May. The European Super Proton Synchrotron (SPS) had arrived. Fermilab's response was swift; on 14 May the new US laboratory took its protons to 500 GeV. The following month, Adams was able to announce to CERN Council that the SPS had accelerated its protons to the design energy of 300 GeV and formally asked Council's permission to go higher. The next day he was able to report that 400 GeV had been reached. However, Fermilab, not content with even 500 GeV, and with new superconducting magnets being developed, added a new name to the machine vocabulary—the Tevatron, aimed at 1 TeV, i.e. 1000 GeV.

Even apart from their physics achievements, the rate at which new projects were appearing on the scene was breathtaking. No sooner had one vast new project made its debut on one side of the world stage than a rival would burst in from another direction and claim attention. To maintain the momentum of the funding bandwagon, older players bowed out as eager new projects jostled in the wings. As well as the UK machines, a synchrotron at Argonne, near Chicago, and eventually the famous Bevatron at Berkeley (subsequently renamed the Bevalac) were phased out. For the future, Brookhaven groomed its ISABELLE (from Intersecting Storage Accelerator) scheme, which could trace its origins back to the Ramsey Panel recommendations of 1963. To be built in a 3.8 km tunnel, ISABELLE would be an intersecting storage ring accelerator, taking 30 GeV protons from the existing AGS, accelerating them in opposite directions in two rings of superconducting magnets and colliding them at energies of up to 200 GeV per beam. A 'Super-ISR', ISABELLE had to grapple with two technologies: colliding proton beams and cryogenics. In October 1978, ground was broken for the new project. Colliding proton beams had been mastered at the ISR and superconducting magnets were being mastered at Fermilab, but ISABELLE and the Tevatron had different superconducting magnet designs. The Tevatron used 'warm iron', with only the current-carrying coil cryogenically cooled. For ISABELLE the preference was for 'cold iron' immersed in a cryostat. Only Fermilab was to succeed initially. For the first time, a major particle accelerator project was to bungle its stage debut, but the magnet design proposed for ISABELLE went on to leave its mark in subsequent superconducting magnet schemes.

VACUUM-PACKED PHYSICS

'To think is difficult. ... To think about nothing is more difficult than about something.'

Lev Okun

With everyday logic becoming increasingly deficient as the complexities of Nature are unfolded, fundamental physics demands many new ingredients. New mathematical techniques are required to keep pace with fresh discoveries and physics insights. Newton, for his theory of gravity, and Einstein, in building relativity, had to develop obscure but powerful mathematics—differential calculus and Riemannian geometry respectively—to provide a framework for their new pictures. Such interrelated advances cannot be guaranteed to happen in the optimal order, and specialists working in one field may not even know that progress is being made from another direction. Fettered by preconceptions and blinded by ignorance, even the most gifted scientists have to blunder their way through the unknown. The overall problem resembles a jigsaw puzzle, which solvers can begin anywhere, and gradually the overall picture emerges. The final solution does not depend on the order in which the pieces are fitted together, but sometimes progress can be blocked by a stubborn piece of the puzzle.

In the physics puzzle, two major pieces had finally interlocked in 1864 when James Clerk Maxwell wrote his historic paper 'The dynamical theory of the electromagnetic field'. For some time physicists had known that electrical and magnetic effects accompany each other (a wire carrying a current produces a magnetic field), but it needed the genius of Maxwell to construct the mathematical envelope. As well as increased understanding, this unification of two apparently different effects, electricity and magnetism, brought its own reward; taken together, Maxwell's equations resembled those of waves, suggesting that electromagnetism could also be carried as a wave, and the existence of such electromagnetic vibrations was dramatically demonstrated by Heinrich Hertz in 1888.

However, waves need something to vibrate in, and this led physicists to suggest a new medium, the ether, supposed to permeate all space, in which electromagnetism worked, but the idea of the ether was short lived. The discovery that light (electromagnetic radiation) moves at a speed which is independent of its source was initially a paradox, but its explanation by Albert Einstein's revolutionary new ideas of relativity were one of the great advances in physics understanding. Although relativity made the concept of the ether redundant, the word had seeped into common vocabulary, and physicists had seen how the vacuum might be put to work.

Electromagnetic waves, such as those of light, are one of Nature's great communicators, transferring messages from one place to another as 'action at a distance'. In the twentieth century, the concept of electromagnetic waves underwent a quantum overhaul, when it was discovered that, at the atomic level, radiation is not a continuous stream of waves but instead looks like rainfall. The quantum 'raindrops' (photons in the case of electromagnetism) are massless messenger particles. Brave attempts to recast electrodynamics in quantum form were made in the mid-1930s. A decade later, new revelations from high-precision microwave spectroscopy led to the Feynman–Schwinger–Tomonaga picture of quantum electrodynamics, the most precise theory known to physics.

Electromagnetism is not alone in Nature's repertoire. With the advent of subatomic physics early in the twentieth century, new effects were discovered, such as nuclear beta decay, a reshuffling of the electric charges of the participating particles. In a simple example of beta decay, a neutron decays into a proton, releasing an electron and a neutrino. This nuclear beta decay appeared inextricably entwined with electric charge. Undeterred by the initial difficulties in formulating quantum electrodynamics, in 1933 Enrico Fermi attempted an ambitious theory of beta-decay electrons and neutrinos, which he first submitted to the journal *Nature*. It was rejected as it contained 'speculations too remote from reality', and Fermi's idea was published instead in Italian.

Whatever it was, the agent of nuclear beta decay was much weaker than electromagnetism, and such effects became collectively known as the 'weak nuclear force'. Whatever carried the weak force also has to be heavy. Electromagnetic action is carried by the massless photon, able to travel over long distances, but the weak force, with its much shorter range, needed a heavy carrier, called by Fermi the W.

Neutrinos are unique among particles. Carrying no electric charge, they feel only this weak force. Physicists hoped that neutrinos would reveal more about how the weak force operates and perhaps even reveal the W. Even though the neutrinos themselves carried no electric charge, each interaction nevertheless continued to show that electric charges are being shuffled around, with a charged particle, a muon or an electron, emerging. There seemed to be no way that the weak force could operate without affecting electric charge. This link between electric charge and beta decay had given several physicists a hunch that these two forms of action at a distance might somehow be connected, but any similarities could not be taken too far; while the weak force carrier appeared to be electrically charged, the photon on the other hand is electrically neutral. Electric charges can influence each other without exchanging any electric charge; the transfer of electromagnetic information, for example through electromagnetic waves, is a 'neutral current'. In contrast, the heavy W is itself a charged particle, giving a 'charged current' for the weak interaction.

A full understanding of the weak force had to await two key discoveries: first, that the proton and neutron constituents of atomic nuclei themselves contained tiny hard constituents, and these could be identified with Gell-Mann's quarks; second, that the weak force transmits neutral current messages as well as charged ones. Both these key discoveries came from the same scientific instrument, the French-built Gargamelle bubble chamber at CERN, filled with

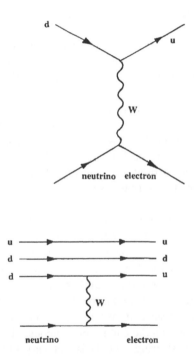

Figure 38: *(a) A down quark (d, electric charge −⅓) and an up quark (u, charge ⅔) are linked by the electrically charged W, which transforms one quark into another. The W also links this quark transformation to an electron and a neutrino. The unit electric charge of the W and the electron matches the difference between the fractional charges of the two subnuclear quarks. (b) Beta decay in terms of quarks. Seen at a nuclear level, one of the down (d) quarks in a neutron transforms into an up (u) quark, while the other two quarks stand by and watch. What begins as a neutron emerges as a proton.*

18 tonnes of Freon. In the early 1970s, Gargamelle's neutrino beam debut changed the course of physics.

In his original 1964 paper introducing the quark idea, Gell-Mann pointed out that heavier quarks would decay into lighter quarks via the weak interaction. A down quark is transformed into an up quark by the electrically charged W. The W links this quark transformation to an electron and a neutrino. The integer electric charge of the emerging electron matches the difference between the fractional charges of the two subnuclear quarks.

Even before the quark picture dispersed the fog obscuring the weak interaction, the similarities between the weak force and electromagnetism had continued to intrigue physicists. Continuing where Fermi and others had left off, Julian Schwinger at Harvard tried to pair the massless photon carrier of electromagnetism with the W Schwinger passed this suggestion to his student, Sheldon Glashow. It was a tough assignment for a young researcher, but Glashow dutifully picked up Schwinger's baton. At first he found it heavy but, as he became accustomed to the burden, he began to make progress.

Electromagnetic and weak interactions look different; making the charged current conform to the weak interactions while the neutral current stayed electromagnetic was difficult. 'One way out,' said Glashow, 'was to introduce an extra neutral current for the weak interactions.' However, then this too had to behave like its charged current companion. With no such neutral current behaviour seen, Glashow had to bide his time.

Theoretical physicists constantly explore fresh ideas, in the same way that a frustrated crossword puzzle solver will reach for a dictionary or thesaurus, or detectives will try to reconstruct a crime or conduct house-to-house enquiries. In the early 1960s, one fashionable such theme was 'spontaneously broken' symmetries, field theories where the overall symmetry is still recognizable, but somewhere the mechanism has been tampered with. A motor car is potentially a symmetric vehicle but, because it has to drive on one side of the road, its steering wheel is offset, giving the car a 'spontaneously broken' symmetry. Such symmetry breaking with steering wheels is not very important when building roads but is vital in designing vehicles. In another example of symmetry breaking, when the temperature drops and initially highly symmetric water starts to freeze, atmospheric precipitation produces snowflakes with beautiful symmetries. Freezing breaks the symmetry of water. In the early 1960s, the implications of such delicate symmetry breaking were being explored in particle physics.

At London's Imperial College the enthusiasm of Abdus Salam was stoking an intellectual furnace. Salam, born in Jhang, in what is now Pakistan, had began work at Cambridge in 1949 and impressed his research supervisor, Paul Matthews, by solving at high speed a problem on generalizing the methods of quantum electrodynamics. Salam had a gift for identifying theoretical ideas whose time was ripe and forcing them to fruition. At Imperial College, theoretical research under Salam was advancing on two fronts. On the one hand a group of research students, including Yuval Ne'eman, was studying the

Figure 39: *Electroweak synthesis pioneer Abdus Salam at Stockholm in 1979, the first Pakistani to receive a Nobel prize.*

symmetries of subnuclear particles, while on the other hand a group was tinkering with spontaneous symmetry breaking. One young researcher who joined eagerly in this work was Steven Weinberg, visiting from MIT.

Weinberg and Glashow went back a long way; they had gone to the same Bronx high school and then to Cornell as undergraduates before setting out on their respective physics careers. By twists and turns, these initially diverging careers were destined to converge at the same historic physics milestone. Glashow

Figure 40: *Electroweak synthesis pioneer Steven Weinberg, seen here at an international conference in 1962.*

remained faithful to the idea implanted by Schwinger. Weinberg, fascinated by the symmetry breaking ideas, tried initially to apply them to quark forces. With the quark ideas so successful, theorists were looking to find what made the inter-quark force work. Ever since Yukawa's suggestion that the nuclear force might have a heavy carrier, experimental physicists had been looking for such particles. With the discovery of the pion, and later its heavier cousins, such as the rho meson, they thought for a time that they had found the keys to the nuclear force, but all attempts to force a result from spontaneous symmetry breaking gave unwanted massless particles. Physicists wanted particles with mass. The only massless particle, so they thought, was the photon. Weinberg was ready to give up working on the spontaneous-symmetry-breaking ideas, quoting Lear's retort to Cordelia, 'Nothing will become of nothing.' Salam and Australian John Ward at Imperial College tried the same route for the weak force. Spontaneous symmetry breaking, with its plague of massless particles, looked to have reached the end of the road.

Help came from another branch of physics, the quantum behaviour of matter in bulk. When atoms come together to form solids, liquids and gases, some component particles lose their affiliation with particular atoms and instead become common property, shared over a volume. Outer orbital electrons are relatively weakly bound, and in a metal or a plasma (a hot gas in which some electrons evaporate from their parent atoms) become quasi-free, gaining a measure of independence. Such communal electrons can vibrate in the bulk matter with characteristic frequencies, in the same way as a sound will resonate in a long pipe. Quantum mechanics says such natural vibrations or waves also show particle properties, having a characteristic mass. In certain cases these quantum properties can also be directional, resembling the rod-like molecules of liquid crystals.

Seeing such effects in other branches of physics, resourceful particle theorists hoped that here could be a way of introducing heavy carrier particles. However, instead of a plasma or a solid, the medium was the vacuum itself, the ether. Instead of atomic electrons, the mobile particles were something else, a property of the vacuum. The possible vibrations of these particles in the vacuum had a preferred direction, as though the vacuum were endowed with a directional 'grain', like a piece of wood. Particles whose symmetry properties made them move along the grain felt no resistance and were light, while those which moved across the grain absorbed energy and became heavy.

One of those adapting these new ideas to particle physics was Peter Higgs of Edinburgh, and ever since this symmetry breaking mechanism has been called after him. Higgs initially submitted his scientific idea to a European physics journal, but it was turned down as being uninteresting. Higgs redrafted the paper, pointing out how the theory could predict new particles, and in 1964 sent it instead to a US journal, where it was accepted. At first the suggestion went unheeded; 'It was just what the doctor ordered to give masses to the (carriers) of the weak force, except that the patient was not awake', wrote Glashow later.

In 1967, Steven Weinberg at MIT took the so-called Higgs mechanism and instead of harnessing it to quark forces turned instead to the unification of electromagnetism and the weak force. Initially he had been '... applying the right ideas to the wrong problem,' he said later. Abdus Salam in London had long nurtured the dream of such a unification, but in the mid-1960s had been preoccupied with quark physics. In 1967, with the Higgs mechanism at hand, he remembered his dream. The ideas of Weinberg and Salam immediately accounted for the electrically charged W particle, the heavy carrier of the weak force, but endowed the W with an electrically neutral partner. This split into two 'directional' components, one massless, corresponded to the electromagnetic photon, and the other heavy, acquiring mass from the texture of the vacuum. According to this picture, the weak interaction should be able to operate without affecting electric charge. With their improved map, Weinberg and Salam were revisiting where Glashow had been before and did not discount the unfamiliar neutral current.

A way of unifying electromagnetism and the weak force had at last been found, but at first few took any notice. With the spotlight firmly on quarks, everywhere else was in darkness and, because of its unfamiliar neutral current, the seeds of the new theory of Weinberg and Salam initially fell on infertile ground. Rather than earning the title 'theory', their proposal was initially described as a 'model'. In physics language, a 'model' is a toy, intellectually instructive or amusing but not necessarily relevant to the real world. As it was initially formulated, the Weinberg–Salam model did not apply to quarks, only to particles like electrons and neutrinos that emerged from beta decay. As well as the drawback of the mysterious neutral current, the model also looked very different from quantum electrodynamics, by which the effectiveness of all other theories was judged. The model did not look as though it would be tractable enough to give well behaved numerical results. In 1971, Gerard 'T Hooft in Utrecht, following a wise suggestion from his teacher, Martinus Veltman, showed that the oddball model, despite its apparent unfamiliarity, did in fact conform to the same sort of rules as quantum electrodynamics and was 'renormalizable' (see page 107). Calculations should be mathematically reliable, even if physics predictions were yet untested. Physicists sat up and took notice. Weinberg, more confident, reviewed previous experiments that had searched for neutral currents and concluded that the rates predicted by his model were low enough to have escaped detection. 'So there was every reason to look harder,' he challenged. 'T Hooft also advocated a further search.

Taking up this gauntlet were two ambitious new experiments using neutrino beams: the Gargamelle bubble chamber at CERN's proton synchrotron, and the first experiment approved for the new multihundred gigaelectronvolt supersynchrotron at the new National Accelerator Laboratory near Chicago, soon to be called Fermilab. Neutrinos, without any electric charge and operating behind the curtain of electromagnetism, would be where any neutral currents should show up. However, in ten years of experiments, neutrinos had always been seen to interact by shuffling around electric charge, with the neutrino transforming into an electron or a muon. In a neutral current process, a neutrino would stay a neutrino but in passing would jolt some other particle into motion. Even without Weinberg's bidding, earlier experiments using neutrino beams at CERN had diligently searched for signs of neutral currents but had found nothing. These initial experiments had not been big enough to offer enough resistance to the neutrinos.

Al Mann from the University of Pennsylvania and David Cline from Wisconsin were eager to have a first crack at a neutrino experiment at the new US accelerator. The first neutrino experiment at Brookhaven had hit the jackpot and there was every reason that higher energy neutrinos would continue this success story. The plan was to build massive spark chambers for a scaled-up version of the classic 1962 neutrino experiment by Lederman, Schwartz and Steinberger at Brookhaven. Carlo Rubbia, a young Italian physicist who had specialized in big electronics experiments at CERN and who had recently moved to Harvard, proposed complementing this spark chamber array with additional electronic counters and improved muon analysis. After formal approval in October 1970 as Experiment 1A (E1A) (the A denoting the addition of the electronic counters) the big detector began to take shape in the Fermilab neutrino area, several kilometres from where the beam would tangentially fly off from the 6.3 km ring. The initial goal of the experiment had been to search for the W particle, the carrier of weak interactions, but its sights were shifted when Weinberg recommended looking for neutral currents. This new aspect of the weak interaction would be carried by an electrically neutral partner for the W—the Z particle.

The two contestants did not look equally matched. The European contender was smaller, while Fermilab also had the advantage in beam energy, 200 GeV protons against CERN's 28 GeV protons, giving neutrinos of some 50 GeV and a few gigaelectronvolts, respectively. It looked in some ways like David against Goliath. With the electron beams at SLAC having just reported evidence for hard structure deep inside the proton, physicists clamoured for Gargamelle's new neutrino's-eye view of protons and neutrons, and the neutral current search did not have first priority. Nevertheless Gargamelle duly started to look for neutral currents in January 1972. Antonino Pullia of CERN suggested looking for neutral current interactions of neutrinos with quarks, subnuclear shakeouts with no redistribution of electric charge—so-called 'muonless events'. This was the direct route to neutral currents, but physicists realized that any such effect would be heavily obscured by interactions by neutrons. Being electrically neutral, neutrons are invisible in a bubble chamber and separating neutron and neutrino effects would require painstaking work. To sidestep this difficulty, another Gargamelle group, headed by Don Cundy at CERN, who had made such searches before using smaller heavy-liquid bubble chambers, preferred to search for processes where the neutrino interacted with an electron. Although these would be rare, such interactions would be 'clean', less liable to corruption by misleading background effects. In 1972, Don Perkins of Oxford, a champion of the electron approach, warned of the difficulties of trying to predict how a neutrino beam would interact with the quark constituents of the nucleon, comparing this approach to the Sultan of Constantinople's seraglio; everyone knew where it was, but few knew exactly what went on inside, even though it was fascinating to speculate. If Weinberg and Salam were right, a few neutral current events should have turned up in Gargamelle's 1971 photo album. None was found. In 1972 a further 360 000 photographs were taken.

In January 1973 at an American Physical Society meeting in New York, Paul Musset of the Gargamelle collaboration presented preliminary results of the continuing search for neutral currents. Theoretical predictions for the neutral current scattering of neutrinos from quarks were becoming more reliable, while the problem of handling the neutron background seemed to be under control, but there were still no neutral currents. However, while Musset was in the USA, the Gargamelle scanning table at Aachen came up with what looked to be a clear example of an electron jolted by a passing neutrino. Helmut Faissner immediately took the photograph to show to a jubilant Perkins at Oxford. The background for the electron route was very low, and the photograph was very convincing. Other physics discoveries, notably the 1964 discovery of the omega minus at Brookhaven, had been made on the basis of a single convincing picture. Confidence within the group grew, and soon the subnuclear route too provided good 'muonless' evidence for neutral currents. The Gargamelle group decided to go public and their claims to have discovered neutral currents, by both techniques, were published in July 1973.

At Fermilab, E1A had seen its first neutrino interactions from 200 GeV protons early in 1973, and by springtime appeared to be seeing plenty of candidate neutral current interactions. Carlo Rubbia was shuttling between the USA and CERN and knew exactly what was happening at Gargamelle. An E1A paper claiming to have seen neutral currents was submitted for publication that summer. Before being accepted for publication, a scientific paper is sent to neutral 'referees' for comment and appraisal. This initial E1A paper was greeted sceptically by its referees, and meanwhile subsequent data looked less convincing. There were other possible explanations on the market for the muonless effects. At an international physics meeting in Aix-en-Provence in September, Salam thought neutral currents were in the bag and described the atmosphere as being 'like a carnival'. In his talk at Aix, Weinberg said, 'It is perhaps premature to conclude from all this that neutral currents have really at last been observed,' adding nevertheless in his conclusion, 'The recent discovery of neutral currents... .'

The referees' objections to the original E1A paper lay unanswered, as by now the new Fermilab accelerator was delivering protons at 400 GeV rather than 200 GeV. The detector had also been modified, with different shielding and with larger spark chambers but was to be tormented by 'punchthrough' (data pollution from subnuclear particles sneaking through shielding that was supposed to stop them). The surviving neutral current signal started to shrink, becoming smaller than the estimated errors. The collaboration began to write a paper claiming that neutral currents did not exist. Rubbia, still shuttling backwards and forwards across the Atlantic, kept CERN posted. Were neutral currents really there? Cynics talked about 'oscillating neutral currents'. With Fermilab's new data apparently contradicting the Gargamelle claim already in print, CERN management, unused to major discoveries, nervously asked the Gargamelle group if they wanted to back down. The European experiment went back over its findings but remained firm. E1A collected more data and in March 1974 submitted their neutral current evidence. By the summer of 1974, neutral currents had come to stay.

It had been an exhausting struggle. The two big collaborations had been pitted against each other, and within each experiment different factions had to resolve their differences. Unused to staking physics claims, CERN management had moments of doubt, but the Gargamelle team, to their credit, never backed down. The scars would take a long time to heal, but a new physics effect had arrived, as fundamental as the discovery 150 years previously that the electric current in a loop of wire behaved as a magnet. Many subsequently claimed that the 1973 neutral current discovery at CERN was of Nobel prize proportions, but André Lagarrigue, the driving force behind Gargamelle, died of a heart attack in 1975. Lagarrigue's successor, Paul Musset, died in a mountaineering accident in 1984. The Nobel prize is not awarded posthumously.

With the discovery of neutral currents, the Weinberg–Salam 'model' was promoted to a fully fledged theory. Mirroring the nineteenth-century invention of the word 'electromagnetism' to reflect the new synthesis of electricity and magnetism, Abdus Salam spoke of a new 'electroweak' picture. Unification is a word frequently used, but some demurred. The unification of electricity and magnetism runs deep; a moving current creates a magnetic field and vice versa, but the weak force is not accompanied by electromagnetic effects. In normal circumstances, the two forces exist independently. Nevertheless there was a deep new link. Electromagnetism is an everyday experience while weak interactions are a laboratory curiosity. The new synthesis spanned a wider range of physics conditions in which conventional electromagnetism would not always be the dominant partner. At higher energies, the 'weak' component of the interaction would become stronger. At high energies, such as the fiery temperatures that had existed soon after the Big Bang, electromagnetism and the weak force, as a unified electroweak force, had been indistinguishable but, as the Universe cooled, it reached a point where the weak interaction suddenly 'froze' out, breaking the underlying symmetry and leaving electromagnetism the dominant partner.

The initial ideas of Weinberg and Salam were limited to electrons, muons and neutrinos and took no account of quarks, but quark transitions, each catalysed by a W particle, were at the heart of the weak interactions. How did quarks fit into this new picture? Glashow's first attempts to unify electromagnetism and the weak force had pre-dated Gell-Mann's introduction of quarks. Trying to explain weak interactions without recourse to quarks was doomed. In particular it was known that weak interactions did not respect the strangeness charge. If a neutral current of the weak force were introduced, surely this would not respect strangeness either. Glashow had initially abandoned neutral currents for this reason but, returning to the weak force after the introduction of quarks, Glashow found new hope. There were three quarks (Gell-Mann's original up, down and strange) but four non-quark particles (the electron, the muon and their respective neutrinos). This three–four lineup looked unbalanced. In addition, once physicists knew that neutral currents exist, the neutral kaon was expected to decay into a pair of muons. It does not. What prevents a particle containing a strange quark from decaying into two muons?

Glashow and Bjorken introduced a fourth quark, called by them 'charm' ('in the sense of a magical device to avert evil') to augment the three standard quarks (up, down and strange). Four quarks, arranged as two pairs, matched the electron, the muon and their respective neutrinos. In 1970, Glashow returned to electroweak unification and the quark picture, working this time with two visitors from Europe, John Iliopoulos and Luciano Maiani. Revamping the charm idea, Glashow, Iliopoulos and Maiani showed that the charm and strange quarks got in each other's way and stopped the neutral kaon decaying into two muons. However, nobody had seen any particles which called for a fourth quark and few took any notice. It was like neutral currents all over again.

With Fermilab delivering its first high energy beams in 1972, the first machine proposed by the 1963 Ramsey Panel had become a reality. Another item on the Ramsey list had been an electron–positron collider at Stanford. Carefully nursed by Burton Richter at SLAC, a modest proposal costing less than US$10 million had initially been left out in the cold after the big money had been diverted for the construction of Fermilab. After several unsuccessful direct approaches for construction money for a new machine, Richter realized that his trump card was the modest size of his project, which could be disguised as a 'facility', a laboratory improvement rather than a new machine in its own right. In May 1971 a modest 'improvement' of 80 m diameter began to take shape on a parking lot in the shadow of the big halls housing detectors for the 2 mile Stanford linear accelerator. Less than two years later, beams were colliding in the Stanford Positron–Electron Asymmetric Rings (SPEAR).

The first thing to do was to carry out a broad survey of the new terrain. That summer, SPEAR had appeared to have seen nothing in a first sweep. The new terrain looked featureless, with each energy step reporting the same response. However, closer inspection revealed a small bump at one energy step. Initially the plan had been to move on to a higher energy band but, in October 1974, collaboration member Roy Schwitters helped to persuade his colleagues not to burn its boats, and SPEAR retraced its steps. At a collision energy of 3.1 GeV (1.55 GeV per beam) the detector suddenly burst into life. Carefully fine-tuning the energy, the SPEAR team homed in on a particle call-sign which almost deafened their detector.

On the other side of the USA, at Brookhaven, Sam Ting of MIT had mounted a big experiment to look for heavy analogues of the electromagnetic photon. Among the gamut of quark–antiquark mesons were several, the rho, the omega (written in lower-case Greek; the capital letter was reserved for the 1964 omega minus which had dramatically confirmed the quark picture), and the phi, which, although subnuclear particles, under certain circumstances behaved like photons and could take part in electromagnetic processes at high energy, providing the photon, hard pressed by being out of its energy depth, with a respite. Ting wanted to see whether there were more of these 'heavy photons' and built a large V-shaped detector to intercept emerging electron–positron pairs. Ting had been keen to build his detector at Fermilab but had been unable to gain a foothold at the new laboratory and had turned instead to Brookhaven. In the summer of 1974 the experiment began to see something in an otherwise smooth electron–positron yield. A few counts began to come in around 3.1 GeV. It could have been a statistical hiccup but, on dividing the electron–positron energy range into narrow slots, all the extra counts seemed to fall into the same slot. Carefully checking it out, by November the group was convinced that the effect was real.

Simultaneously, the teams led by Richter at SPEAR and Ting at Brookhaven had contrived to discover a new particle at 3.1 GeV and in opposite ways. SPEAR had synthesized the particle by forcing electrons and positrons together under the right conditions. Ting had seen it by its decay into electrons and positrons. Reflecting its dual discovery, it was given two names, J by Ting and psi by Richter, and ever since has been known as the 'J/psi'.

The discovery of the J/psi—'the November Revolution'—caught most people by surprise. The first issue of *Physical Review Letters* for 1975 contained eight explanations for the J/psi, of which two turned out to be correct, one by Thomas Appelquist and David Gross, and the other by Alvaro de Rùjula and Sheldon Glashow, all at Harvard, who said it was due to the fourth quark, 'charm'. The J/psi was made up of a charmed quark and a charmed antiquark locked together. The quark and antiquark charm mutually cancel, making the charm difficult to see. However other particles whose quark configuration included a single example of charm soon turned up. For their spectacular discovery, Richter and Ting shared the Nobel prize for physics in 1976.

After Richter's team had found the psi, the energy at SPEAR was inched higher. Martin Perl, who had been Sam Ting's research supervisor at Michigan, was looking for new particles at the new collider, but not of the J/psi kind. Using the same detector as Richter's team, Perl intercepted a handful of U (for unknown) events which suggested a new particle. In the excitement of the spectacular J/psi discovery, Perl's U events were initially overlooked, but Perl was eventually to win the 1995 Nobel prize for his discovery of a third weakly interacting particle, the 1.8 GeV tau, a partner for the electron and the muon. This low-key discovery also suggested that four quarks were not the end of the story. A third weakly interacting particle needed two more quarks to partner it.

Electron–positron colliders had been built in Europe and Russia, but it had been SPEAR in the USA that bet on the correct energy, providing a bonanza of new particles. At these energies, electron–positron colliders were compact economical machines which did not need as much political and financial underwriting as the big proton synchrotrons. Richter's SPEAR showed how physics could be done relatively cheaply. After SPEAR, the next major collider to appear on the scene was at the Deutsches Electron Synchrotron (DESY) in Hamburg.

While Germany was one of the founder member states signing the CERN convention in 1954, some German physicists felt that their country's distinguished scientific tradition merited an additional commitment in subnuclear physics. With CERN initially dedicated to protons, Germany's national effort took instead the electron route. Wolfgang Paul in Bonn built an electron synchrotron in 1954, the first European machine to use the strong-focusing principle. At an accelerator conference in Geneva in 1956, a group of German physicists met at the home of Wolfgang Gentner, then working at CERN, to look at the idea of building a 6 GeV electron synchrotron. Willibald Jentschke, just returned from the USA to head the Hamburg Institute of Physics, was elected to lead the new project. Eventually reaching 7.2 GeV, the DESY electron machine began operation in 1964. Until the arrival of the 2 mile Stanford linac, it provided the world's highest-energy electron beams. While national centres in the UK and France had opted to build proton machines which were subsequently eclipsed by CERN, DESY's choice of the electron route provided a useful physics alternative to Europe's high-energy proton diet.

For European scientists, working in the USA still had glamour and attraction, while the cosmopolitan CERN environment had its obvious attractions, including the world's best physics cafeteria. DESY, in its remote leafy Hamburg suburb, had less going for it, except for those who wanted to work with electrons. Sam Ting had been an early customer. With its electron synchrotron, DESY had planned its own electron–positron collider, DORIS (DOppel-RIng-Speicher), an oval 50 m by 100 m, which had come into operation slightly too late to catch the J/psi action, although DORIS was soon providing more candidates for charm quark–antiquark bound states. DORIS also confirmed Perl's tau particle.

With the electron–positron route looking to provide such good research returns on investment, laboratories which had traditionally relied on electrons looked towards a new generation of larger electron–positron colliders. Cornell, with Robert Wilson's electron synchrotron, SLAC in the USA and DESY in Europe prepared plans for more ambitious electron–positron rings. This next generation of colliders, CESR (Cornell Electron Storage Ring), PEP (from an original Proton–Electron–Positron plan) at Stanford, and PETRA (Positron Electron Tandem Ring Accelerator) at DESY, hoped for rich quark spectroscopy pickings at higher energies.

However, Sam Ting's double-armed detector at Brookhaven had shown that proton machines could also play this game. Fermilab, with the world's highest-energy proton machine, and irritated by having lost the race to neutral currents, was eager to stake a major discovery claim. It duly came in the summer of 1977, when a team led by Leon Lederman paralleled at higher energies what Sam Ting had done at Brookhaven three years before. Lederman's team was rewarded by a sharp resonance at 9.5 GeV, the upsilon. It was the first sighting of a fifth quark, 'beauty' (b), to partner Perl's tau. Just as the J/psi was composed of a charmed quark and antiquark stuck together, so Lederman's upsilon was a beauty quark locked with its antiquark. More details of upsilon spectroscopy, together with sightings of other particles containing b quarks, soon came from DORIS, boosted with additional radio-frequency power to reach the upsilon threshold, and the new CESR ring at Cornell, while Novosibirsk provided a precision measurement of the upsilon mass.

Even while DORIS was straining to reach the upsilon, DESY was pushing ahead with the construction of a electron–positron ring, PETRA, of 730 metre diameter. Under the firm guidance of DESY Director General Herwig Schopper and a machine team led by Gustav-Adolf Voss, who had learned the accelerator business in the USA, PETRA began operations nine months before the planned date, and two years before a comparable ring, PEP, at SLAC. PETRA's big payoff came in 1979, the year before PEP came into operation. The flashes of energy produced by annihilating electrons and positrons create quark–antiquark pairs bound together, like the J/psi or the upsilon. Surplus energy is absorbed by the quark–antiquark system, and eventually the gluon mesh joining the quark and the antiquark snaps. In the same way as cutting a bar magnet in two creates fresh magnetic poles on the cut ends, so the two pieces of gluon mesh have appropriate quarks on the ends. The waiting detectors see two 'jets', tightly confined sprays of subnuclear particles, emerging in opposite directions.

However, three theorists at CERN, John Ellis, Marie Gaillard and Graham Ross, pointed out that the quark–antiquark system could also lose some energy in the form of a gluon. As well as two sprays of subnuclear particle energy emerging from the quark–antiquark system as back-to-back jets, there could also be three-jet patterns, looking something like the Mercedes–Benz trademark. After developing careful techniques to ensure that the right particles were assigned to the right jets, in 1979 the experiments at PETRA saw clear signs of such a star pattern. The veil had been lifted from the gluon, the carrier of colour force between quarks, long thought to be permanently locked into the inter-quark force. Ten years after SLAC's electron beams had first revealed the inner structure of the proton, here was explicit evidence for the particle which holds the proton together and carries some 50% of its energy.

Having mastered the tricky problems of labelling jets, it turned out to be less easy to say which of the four PETRA detector teams, JADE, Mark-J, PLUTO and TASSO, had actually made the discovery. Most people thought that TASSO had got there first, but Sam Ting at Mark-J had a different way of doing things and had also staked a claim. The dilemma was disinterred in 1995, when a prestigious prize was awarded by the European Physical Society (EPS) to Paul Söding, Björn Wiik, Günter Wolf and Sau-Lan Wu, key members of the TASSO group, for their three-jet work. In an unprecedented move, the EPS Executive

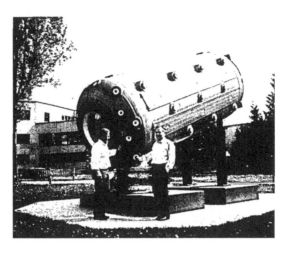

Figure 41: *After its epic discoveries at CERN, the Gargamelle bubble chamber Gargamelle retired from physics in 1978 after it developed a structural fault. The once-proud detector is now an open-air exhibit at CERN's visitor centre, where curious tourists contemplate its enormity.*

Committee also awarded a special complementary prize to all four PETRA collaborations 'for establishing the existence of the gluon in independent and simultaneous ways'.

However, nobody thought of giving a prize to Gargamelle. After its epic discoveries at CERN's proton synchrotron, in 1976 Gargamelle was laboriously moved to receive higher energy neutrino beams from CERN's new SPS, but it was as if this sudden extra effort, coming so soon after such a prolific period of discovery, was too much. In 1978 a structural fault was found in the body of the chamber and Gargamelle retired from physics. The once-proud bubble chamber is now an open-air exhibit in the park adjoining CERN's visitor centre, where curious tourists contemplate its enormity, but its real monument is in the indelible mark it has left on science. In the space of a few years in the early 1970s, Gargamelle's discoveries provided the key which opened the door to our understanding of weak interactions.

PHYSICS FROM AA TO Z

The Solomonic wisdom of siting CERN's new SPS next to the existing proton machines had its price. With Member States having vied for so long to host the big new machine, a whole new administrative scaffolding had been erected, and this creaky edifice did not disappear overnight. For five years from 1971, the eccentric legacy of the turbulent SPS negotiations was two semi-autonomous but uncomfortably close CERNs: the 'old' site, CERN I, and a new CERN II for the SPS.

Amid this schizophrenia, Gargamelle at CERN I discovered the elusive neutral current of the weak interaction, and the electroweak unification of the electromagnetic and weak forces looked to be in business. To clinch the new theory, physicists had to find the key particles which carried the charged and neutral currents in the new electroweak picture, universally called the W and the Z, respectively. Few imagined that CERN II's new SPS was to be the stage for their discovery.

Rather than finding the missing W and Z, the primary concern of CERN Council in 1975 was instead to remove the embarrassment of having two CERNs. Adams' mandate as Director General of CERN II still had several years to run. A new Director General could be appointed for CERN I and the two laboratories could continue to coexist for another three years before being merged, but this would only prolong the agony. For an immediate merger, who would run the combined laboratory? Front runner for the new CERN I Director General post was Leon Van Hove, a gifted Belgian physicist who headed CERN's Theory Division and was looking to spread his wings. With various options proposed, the solution finally adopted was to bring the two laboratories together from 1976, but under two Directors General: Van Hove as Research Director General to look after the research programme, and Adams as Executive Director General, with responsibility for managing the laboratory.

With the new SPS coming into operation, it should have been a proud moment for the enlarged European laboratory and its gleaming new synchrotron, but there was an uneasy feeling that the SPS had been beaten to the punch by Fermilab, which had already been running for several years and at a higher energy. In addition, the successes at the modest SPEAR electron–positron collider at SLAC had led to a rash of electron–positron colliders springing up everywhere.

Even though the SPS had only just started to do the job for which it had been designed, it was already time to start thinking about CERN's next major

143

research project. There was no lack of ideas on the table: higher energies in the ISR, colliding electrons and protons in the SPS, and, an initial favourite of Adams, to emulate Fermilab, equipping the SPS with superconducting magnets to take it to much higher energies. All of them had their proponents and some sales talk, but none had a really convincing physics case. The overriding physics objective was to find the W and Z particles. New neutrino experiments, including those at the SPS, were showing exactly how the neutral current and the photon were related, and physicists had a good idea of the masses of the W and Z carrier particles—slightly under 100 GeV, about the mass of a nucleus of niobium. Making such heavy particles leave their usual carrier role and come out into the open was beyond the reach of any existing machine.

At a sabbatical year at CERN from 1975 to 1976, Burt Richter was one of a team looking at the possibility of a large (about 30 km) electron–positron ring specifically to hunt for the Z carrier particle. Electrons and positrons annihilate to provide energy and, if the energy were tuned to the neutral Z, then this ring would mass-produce them. If the energy could be cranked high enough, pairs of oppositely charged W particles would also be attainable with LEP (Large Electron–Positron Ring) but subsequently abbreviated to LEP). LEP emerged in the traditional ECFA European forum as a leading contender for the next major project.

Meanwhile Carlo Rubbia and David Cline, who had worked together on the first big Fermilab neutrino experiment, together with Peter McIntyre and Fred Mills at Fermilab had another scheme. The neutrino experiment at Fermilab, had been proposed as a search for the W but had turned to the search for neutral currents instead. The W goal was still there. The Cline–McIntyre–Mills–Rubbia idea was to take protons and their antimatter counterparts, antiprotons, and to do with them what the collider specialists had done so far only with electrons and positrons. Take a big proton synchrotron and fill it with high-energy protons. Then get hold of some antiprotons and put them in the synchrotron too, where they would circulate in the opposite direction to the protons. With the synchrotron operating instead as a storage ring, the two beams would be smashed together. The electron–positron colliders then being prepared at DESY, Cornell and SLAC were aiming at collision energies of around 10 GeV. With beam energies of several hundred gigaelectronvolts, the proton–antiproton scheme could in principle provide collision energies of almost 1000 GeV, enough to make W and Z particles come flying out. Continuing his transatlantic commuting, Carlo Rubbia introduced the idea to Fermilab and to CERN, both with new multihundred gigaelectronvolt proton synchrotrons.

Ideas for a scheme to collide protons and antiprotons had been prepared earlier at Novosibirsk's Institute of Nuclear Physics, under the imaginative guidance of Gersh Budker. A plan for a proton–antiproton scheme, VAPP–NAP, with collision energies of up to 45 GeV had been described at an international accelerator conference at CERN in 1971. Budker knew that feeding antiprotons into a storage ring would be difficult. Antiproton beams had been made for experiments, but such antiproton beams were haphazard affairs, with particles going all over the place. Filling a synchrotron that way would be like trying to feed a hosepipe from a shower attachment.

Electron beams are naturally well behaved because electrons are light particles. If a high-energy electron beam is bent by a magnetic field, the light electrons 'skid', emitting a high-pitched screech of radiation, which saps energy from the beam which would otherwise make it unruly. Heavy particles such as protons do not lose energy so easily as they bend round corners. To tame unruly proton beams, something has to be done. Physicists call the removal of such unwanted beam energy 'beam cooling'. At Novosibirsk, the talented Gersh Budker had shown how passing unruly particles through a narrow 'sleeve' of orderly electrons could absorb unwanted sideways momentum. At CERN, Simon van der Meer, who had invented the 'magnetic horn' to improve neutrino intensity and who more recently had been in charge of the power supplies for the SPS, had developed an alternative scheme for beams circulating in a storage ring. In van der Meer's 'stochastic cooling', the idea was to measure the 'centre of gravity' of the circulating beam at one point and to calculate its departure from the ideal value. The required correction signal would then be fed diametrically across the ring to a 'kicker' on the far side, ready to nudge the wayward centre of gravity as it came round. Even if the cooling systems worked and the antiprotons could be tamed, it would mean that the proton synchrotron would have to spend time being something other than a proton synchrotron, no longer catering for experiments anxiously waiting downstream for their particle beams.

Nevertheless, at both CERN and Fermilab the idea had its attractions, but Fermilab was more committed to its superconducting Tevatron to complement its original synchrotron Main Ring, which opened several collision options: protons on protons, protons on antiprotons, Tevatron–Tevatron and Tevatron–Main Ring. However, with Fermilab seeing the Tevatron as its primary long term goal, collision physics was relegated to second place. At CERN several collision routes beckoned: a higher energy version of the ISR, this time fed by the SPS, for proton–proton collisions, protons and antiprotons in the existing ISR, and protons and antiprotons in the SPS. In addition, there was the major long-term goal of the large LEP electron–positron ring. CERN, which was by now less enthusiastic about emulating Fermilab with a superconducting SPS, saw collision physics as a primary goal, in parallel with LEP. With most of the infrastructure in place, the collider would be on a fast track, with LEP as a longer term objective.

Leon Van Hove, moving into his new Research Director General driving seat, pushed for a bold research line. Providing research facilities for thousands of physicists from all over Europe and even farther afield, CERN's heavy committee structure and continental European correctness had preferred to back 'safe' experiments, leaving the more flamboyant Americans to run risks, but also letting them walk off with the Nobel prizes. At first, getting international collaboration to work had been recompense enough for CERN, but this research had yet to reap rewards. Media comment that CERN consistently missed

Figure 42: *In 1976, the new Research Director General Leon Van Hove pushed for a bold CERN research line.*

headline discoveries rubbed salt into the wound. Piqued by this lack of apparent success, Van Hove seized on the Rubbia proton–antiproton proposal and convinced Adams of the project's worth. With unified backing, it was pushed through as an integral part of CERN's existing programme.

It was a bold decision. 'Too bold', some had said. The SPS was only just getting into its stride, and it would be a pity to have to close it for modifications, even if only briefly. Proton–antiproton technology was only just technologically feasible. Even if all the new techniques worked, throwing three quarks and three antiquarks at each other at energies never reached before could produce a mass of confused debris, masking any W and Z particles.

To test the feasibility of such a scheme, and to compare the relative merits of

Budker's electron cooling and van der Meer's stochastic cooling schemes, CERN built a small test ring, the Initial Cooling Experiment (ICE). By 1978, van der Meer's stochastic cooling route seemed a better option under CERN conditions for taming unruly beams than was Budker's electron cooling, but the ICE work had been done with protons, and there was one final test to make. Protons are stable, and so, in principle, are antiprotons. However, nobody had ever been able to test this. Perhaps antiprotons, like the heavy particles revealed in high-energy experiments, eventually disintegrated. The longest that a single antiproton had been in view was 140 microseconds. There would be no point in building elaborate and expensive antiproton facilities if the antiprotons evaporated before anything could be made to happen. In 1979 the ICE ring received its first taste of antiprotons and the physicists held their breath. The antiparticles coasted happily for hours. CERN gave the green light to Rubbia's proposal, and colliding protons and antiprotons became a challenge instead of a dream.

The heart of the new CERN antiproton project was a specially built antiproton 'factory' in which proton antimatter would be mass-produced and forged ready for its new role. Pulses of raw antiprotons, magnetically selected from the produce of protons dumped into a target, had to be fed into a new 'Antiproton Accumulator' ring, there to be tamed with van der Meer's stochastic cooling. Once tamed, each antiproton pulse had to be switched to a second 'stacking'

Figure 43: *Carlo Rubbia, whose idea to find the W and Z particles in proton–antiproton collisions was one of the great physics sagas of the twentieth century.*

orbit alongside, and another antiproton pulse fed in, cooled and transferred to the stack. After several days and several hundred thousand injected antiproton pulses, a million million antiprotons should be orbiting in the stack, a match for a normal proton beam. At this point, the antiproton stack would be tipped out of the Antiproton Accumulator and the energy boosted, first in the proton synchrotron, and then in the SPS, in tandem with contra-rotating proton beams. When the proton and antiprotons reached their final 'coasting energy' in the SPS, the two beams would be nudged into collision.

On 3 July 1980, the Antiproton Accumulator was put through its paces in a dry run with protons. With everything working as it should, the currents in the electromagnets switched direction, and the first antiprotons were fed in. However, with only a few stochastic cooling units commissioned, initial antiproton intensities were a few per cent of what was eventually hoped for. Meanwhile new tunnels were being completed to feed the antiprotons from one ring to the next. By the summer of 1981, all the CERN accelerators had been run in for antiprotons and, only three years after the project was authorized, the first high-energy proton–antiproton collisions were recorded. The new production

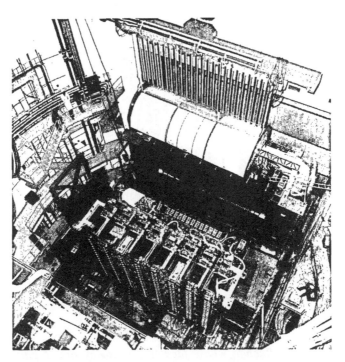

Figure 44: *Carlo Rubbia's UA1 detector at CERN's proton–antiproton collider, shown partially dismantled during maintenance.*

line was in business. Carlo Rubbia delayed his departure to an international conference in Lisbon so that he would be able to announce the news. CERN's gamble looked to be on the right track and the emphasis shifted from the machines providing the antiprotons to the experiments recording and analysing the resulting proton–antiproton collisions. Would the W and Z particles deign to show themselves?

To look for these collisions, huge detectors had been constructed, thousands of tons of equipment arranged in concentric boxes surrounding the proton–antiproton collision point, like a huge high technology Russian doll. Each box was designed to capture one aspect of the collisions and, by putting the information from all the boxes together, the physicists would arrive at a complete picture. These detectors were installed in huge underground cathedrals around the collision points in the CERN SPS. Operating these huge detectors needed teamwork on a scale never seen before in a scientific experiment, with some 200 physicists involved. Subsequent efforts were to be much bigger, but the proton–antiproton experiments at CERN in the early 1980s set a new scale in scientific collaboration. Responsibility for the different detector components was shared out between the collaborating research institutes and universities. Hundreds of man–years of effort went into the design, assembly and testing of the thousands of units for the detector subassemblies. Wire by wire, module by module, the complicated electronics were put together and piece by piece the detectors took shape in the participating institutes. As each detector element passed its acceptance tests, it was shipped to Geneva. The logistics of the operation were immense, and the final dimensions of the detectors were in some cases determined by the maximum size of equipment which could be transported to Geneva.

Looking at their first proton–antiproton collisions, physicists saw clear signs of the tightly confined sprays of emerging particles, 'jets', which showed that the quarks and antiquarks deep inside the colliding protons and antiprotons were actually hitting each other. Quark jets had never been seen so clearly, but the physicists knew that it was too early to look for the long-awaited W and Z particles. Although the energy was there, the proton-antiproton collision rate was still too low.

By 1982 the collision rates had been coaxed higher, and the curtain was ready to go up on the first proton–antiproton run with real chances of seeing W and Z particles. However, an accident in Rubbia's detector meant that carefully prepared components had to be taken apart and recleaned, and the run had to be postponed. Physicists grumbled, but the delay turned out to be a blessing in disguise. Instead of a series of short runs interspersed with other SPS physics, the collider programme was compressed into one long run later in the year. It took time to get an antiproton act together, and continuity was not improved by chopping up the show. However, with one long run, the accelerator specialists were able to perfect their new skills and to ensure that their antiproton stacks were more reliable.

In the summer of 1982, British Prime Minister Margaret Thatcher, on holiday in Switzerland, paid a private visit to CERN. With her background as a chemist, she had always taken a strong interest in basic research. During her visit, she asked CERN's Director General Herwig Schopper, formerly the leader of the German DESY laboratory, to let her know immediately when the W and Z particles were found. She did not want to have to rely on press reports.

A few months after Margaret Thatcher's visit, the proton–antiproton collision score had been boosted a hundredfold compared with the previous year. The circulating antiproton beams became more and more durable, one antiproton shot being held captive for almost two days before it finally frayed and had to be ditched in favour of a new shot. By this time the experiments had seen several thousand million proton–antiproton collisions. Knowing exactly what they were looking for, the experimenters had prepared digital expressways to shovel the data through their computers. Pre-programmed electronic turnstiles were set to click each time a collision with a valid W ticket arrived. Although nobody said much at first, some satisfied smiles were being seen around the experiments' control rooms. By December, Schopper was confident enough to forewarn Margaret Thatcher. Early in the new year, ten candidate W particles could be held up for inspection but, with the implications of the discovery so momentous, no scientific claim was yet made. Over the weekend of 22–23 January 1983, Carlo Rubbia became more and more convinced, 'They look like Ws; they feel like Ws, they must be Ws,' he remarked. The announcement that the electrically charged W had been seen was made on 25 January.

The next goal was the Z, the W's electrically neutral companion. The experimenters knew that the Z needed more proton–antiproton collisions than the W, but on the other hand its 'fingerprint' should be easier to spot. In April 1983 a new antiproton run began. Although more antiprotons were being formed, collected and accelerated, the Z did not seem keen to show itself but, as May went by, the first Z calling cards were seen. On 1 June the Z announcement was made. The carriers of the weak nuclear force had been found. Physicists called them 'intermediate bosons', but more imaginative press reports called it the discovery of 'heavy light', alluding to the deep link between the new particles and electromagnetic radiation. For the discovery, Carlo Rubbia and Simon van der Meer were awarded the Nobel physics prize in 1984, one of shortest intervals in Nobel history between a discovery and the Nobel award.

Going for the W and Z had been a scientific and technological risk. Sheldon Glashow, Abdus Salam and Steven Weinberg, the three theorists who had pieced together the electroweak theory, had been awarded the Nobel prize in 1979, a courageous move by the Royal Swedish Academy of Sciences in view of the fact that the key W and Z particles predicted by their theory had not yet been seen. As one physicist commented, 'Does this mean they will have to give back the prize if the W and Z are not found?' The question remained hypothetical. While there was no doubt that many more elementary particles had been discovered in America than in Europe, CERN's new discovery curiously meant that all the carrier particles (the photon of electromagnetism, the gluon of the inter-quark force and the W and Z of the weak force) were seen for the first time in Europe.

Figure 45: *The vision of Bob Wilson: Fermilab, on the Illinois plain near Chicago. A portion of the 4 mile ring is visible on the right. In 1995, experiments at Fermilab discovered the sixth quark: 'top'.*

While CERN was busy with W and Z particles, Fermilab completed its superconducting Tevatron in 1983. Initial running was at 800 GeV per beam and was later increased to 900 GeV. With the Tevatron complete, the US laboratory could at last turn to its shelved proton–antiproton project. Fermilab's Tevatron collider started to operate in 1985 and had the advantage of beam energies several times higher than those of CERN. The Fermilab collider's big moment came in March 1995, when its experiments discovered the sixth quark, by far the heaviest of all, the 'top'.

With six quarks on the shelf, subnuclear particles and their interactions could be repackaged. The particles fell into two kinds: quarks, which feel the colour force of gluons, and the weakly interacting particles (leptons) which do not feel the colour force. Quarks came in pairs, each pair teaming with a pair of leptons to make a quark–lepton 'family'. Thus the up and down quarks (the constituents of ordinary nuclear matter) pair with the electron (the constituent of atoms) and its neutrino. The strange and charmed quark pair with the muon and its neutrino. In the third family come the beauty and top quarks and Perl's tau lepton and its neutrino. This packaging of quarks and leptons is unimaginatively called the 'Standard Model'.

With its two forms of motive power (the inter-quark colour force carried by gluons, and the electroweak force carried by photons, W and Z particles), the Standard Model can be compared with a submarine. A conventional submarine has two separate sources of power, a diesel engine for travelling on the surface, and an electric motor for use when submerged. In the same way as the submarine's diesel engine and electric motor operate independently, so the Standard Model's two physics 'engines' are decoupled from each other. It is the dream of many physicists to unify the colour and electroweak forces, to provide a single source of physics motive power, a 'Grand Unified Theory', the physics equivalent of a nuclear-powered submarine which can travel anywhere with a single power plant.

The Standard Model had three quark–lepton families, but the idea of particle families had a familiar ring to it. Was there not a risk that this ladder of families would continue for ever? Higher-energy beams could reveal more and more unstable particles, containing yet more quarks and leptons.

While the SPS proton–antiproton collider had still been searching for the W and Z, CERN's LEP electron–positron collider had been prepared as CERN's next major project. With no guarantee that the proton–antiproton gamble would succeed, LEP would provide Z particles, and W particles, too, if the energy could be cranked high enough. If the proton–antiproton gamble did succeed, then LEP would provide the precision physics to follow up the discovery—a no-lose situation. Steering committees and study groups converged on an initial design, seen as a 31 km ring, with its own injection system, calling among other things for a 22 GeV electron synchrotron. John Adams had initially been reluctant to consider a new electron route for CERN, having worked with protons for so long but, when he did add his weight behind the project, he wisely supported the suggestion that the tunnel for the new electron machine be made large enough to leave room for a subsequent ring of bulky superconducting magnets to handle a high-energy proton beam, the so-called Super Proton Electron Complex (SPEC).

Figure 46: *CERN's 27 km LEP ring—the world's largest accelerator.*

From the outset, it was clear that LEP would be built next to CERN. There was no talk of CERN III. As the SPS had already discovered, the terrain around CERN is far from flat, and for a 31 km ring this was even more of a problem. The largest obstacle was the Jura mountain range to the west, rising to a height of some 1700 m. To minimize the portion of LEP tunnel which would have to be blasted out deep under the mountains, the circumference of the ring was reduced to 27 km, with the plane of the ring on a slant, pointing upwards towards the

Jura. Another major modification came with the injection scheme. Instead of building a big new synchrotron to feed LEP with electrons, a way was seen to convert the existing proton machines, fed by a suitable electron and positron source. (This was not the first time that the CERN proton machines had been considered for an additional electron option. In the mid-1970s a plan to add an electron ring to the SPS to give an electron–proton collider had been promoted by UK theorist Christopher Llewellyn Smith and Björn Wiik, a member of the TASSO team at the PETRA electron–positron collider at DESY Hamburg, which was soon to be chasing gluon jets.) In June 1980, CERN Council gave this redesigned LEP the go-ahead. Construction work soon began for the new electron pre-injector infrastructure, on the site where the Gargamelle bubble chamber had stood for its landmark experiments in the early 1970s. Fitting the new pre-injector into the available space was difficult; one of the major roads on the site had to be relegated to a one-way street, with a 300 m detour for incoming CERN traffic.

However, CERN Director General Herwig Schopper's task of building a machine as big as LEP in the increasingly cost-conscious 1980s demanded sacrifices. Unlike the SPS, which had been funded as a separate project, no new money would be available to build LEP. Finding some 1000 million Swiss francs to build the giant new electron–positron collider from CERN's normal operating budget would mean making sacrifices. CERN was being asked to tighten its belt. The ISR were closed prematurely in 1984, described as an act of 'scientific vandalism' by Erwin Gabathuler of Liverpool University. Although the ISR's physics mission was not yet complete, CERN's machine specialists had learned much from the challenge of building and operating the ISR and this experience was to stand them in good stead later. The ISR was more a sacrificial lamb than an act of vandalism.

While CERN's SPS proton–antiproton collider initially had the field to itself, LEP had competition from the outset from SLAC, the major US electron laboratory. Burton Richter at SLAC was one of those who had done the initial calculations for LEP and understood how big electron storage rings could go. Electron synchrotrons or storage rings have to be made large in order to limit the energy losses as the electrons (and positrons) are made to bend round in a circle. LEP, with a diameter of 27 km, would be taking its electrons and positrons initially to about 50 GeV per beam, while the 7 km SPS ring takes protons to 450 GeV. Scaling up LEP to get 500 GeV electrons would mean a ring with a circumference of some 700 km. Building such a machine in California would take up most of the east-west width of the state. Richter, who succeeded Wolfgang Panofsky as Director of SLAC in 1984, concluded that LEP would remain the largest electron storage ring ever built. To get to the high energy electron–positron frontier, SLAC chose another route. Instead of having counter-circulating electron and positron beams in the same ring, the idea was to have two linear accelerators firing at each other, like two mighty cannons whose shells would collide in mid-trajectory. Already equipped with one 2 mile linac, SLAC did not want to build another for positrons, so the plan was to accelerate both the electrons and the positrons down the 2 mile linac, to separate the beams as they emerged from the end of the cannon, and to lead them round 2.3 km semicircular underground arcs to converge at a collision point. The resulting arrangement looks like a giant banjo, the neck being the 2 mile linac. To get at the Z, extra power had to be provided to boost the linac's energy towards 50 GeV.

When two beams of particles collide head on, most of the beams are in fact empty space, so that the particles slip past each other, like ships in the night. The chances of collision are improved by squeezing the beams into as small a ray as possible. A storage ring does this by compressing the circulating beams just as they enter the collision zone, but these circulating beams cannot be made too small, as the interaction between the two beams makes them get in each other's way and become unstable. With a storage ring needing to keep its particles circulating for as long as possible, this limits the size of its beams. However, a linear collider does not have to worry about this as its beams are discarded once they have been through the collision zone.

SLAC's new machine, the Stanford Linear Collider (SLC), began operations in 1989. While the electron and positron bunches in LEP are thin flat slabs several millimetres wide, those in the SLC are only a few microns (a twentieth of a hair's breadth) across. Even pinpoint accuracy is not good enough if two such beams have to be locked together in head-on collision, and the 'final focus' requires special precision work.

LEP's first circulating beams had been announced for 14 July 1989, the two-hundreth anniversary of the French revolution. Knowing that Z particles would pour out of LEP, the other machines strained for a share of the Z action before they would be overshadowed. On 21 June, initial results came from a hundred Z particles netted by SLC. By the time of a big international physics meeting, held at Stanford that summer, the SLC Z catch had climbed to over 200. Fermilab's proton–antiproton collider had thrown 600 more into the kitty.

The width of a particle resonance, the energy range over which it can be detected, is a measure of its stability. A narrow spike resonance with no width at all is infinitely stable. The more unstable it becomes, the wider is the resonance. A resonance can be compared with a bucket; if the bucket is watertight, the contents will not be lost. However, if holes are drilled in the bucket, its contents are lost; the more holes, the faster the water comes out. The width of the resonance is a measure of how many 'holes' there are in the bucket. The proton–antiproton colliders produce Z particles from quark–antiquark collisions deep inside the protons and antiprotons. With the individual quark energies difficult to measure, proton–antiproton colliders find it hard to see the Z as a resonance with a width. However, electron–positron colliders produce their Z particles directly, the beam energy sweeping across the production resonance. Measuring the width of the Z resonance in this way would at last give physicists an indication of how many different quark–lepton families there were. Three were known, but these could just have been the first three rungs of a long ladder. From the resonance shape of the first 200 SLAC Z particles, physicists were able to deduce that the number of quark–lepton families is 3 ± 0.9. With 350 Z particles in the bag a few weeks later, this limit had improved to 2.7 ± 0.7, meaning that the chance of a fourth quark–lepton generation was less than 5%.

Meanwhile LEP had seen its first beams as scheduled on 14 July, and one month later the four big experiments were seeing their first Z particles. By 13 October, well over 10 000 Z particles had been caught, and the chances of a fourth quark–lepton family had been reduced to one in a thousand, a chance few people, particularly physicists, would be willing to wager on. As divine punishment for daring to compete with LEP, a major earthquake measuring 7.1 on the Richter (not SLAC Director Richter!) scale rocked the San Francisco Bay area on 16 October and put the precision SLC beams out of kilter. It was three months before SLC was back in action.

For the next six years, LEP delivered almost five million Z particles, turning Z physics into a precision science, measuring Z parameters to within a few parts in a hundred thousand. In this exacting work, the 27 km LEP ring became one of the world's largest precision instruments. The behaviour of the electron and positron beam closely reflect any movement in the tunnel itself, where earth tides change the circumference of the ring by about a millimetre twice in every 24 hours. These tidal effects are amplified by the machine and have to be routinely allowed for. With more such perturbations identified and allowed for, precision increased in 1995 towards the one part per hundred thousand level. In the summer of 1995, unexplained blips in the beam energy started to show up. At a loss to explain this new effect, CERN consulted Geneva electricity supply engineers, who immediately showed how the blips were due to current leakage from high-speed electric trains passing nearby, an explanation dramatically underlined by a total absence of blips during a one-day rail strike! Equipped with superconducting radio-frequency equipment to push the electron and positron beam energies higher, in 1995 LEP finally left the Z resonance, and in 1996 set its sights on producing pairs of W particles.

LEP's precision probing of Standard Model physics failed to reveal any flaws. Although many parameters of the Standard Model are arbitrary, they are nevertheless interconnected if the model is to be self-consistent. Like a framework which becomes increasingly rigid as more struts are added, so the Standard Model's parameters have less remaining room to manoeuvre as precision results from LEP and elsewhere pile up. In an impressive demonstration of Standard Model cohesiveness, the mass of the sixth 'top' quark, discovered at Fermilab's Tevatron, was correctly predicted to be around 170 GeV. The Standard Model picture was rock solid, with everything fitting together perfectly: quarks, leptons, and carrier particles. The missing link was the mysterious spontaneous symmetry breaking, namely the Higgs mechanism, which gave the vacuum a preferred direction. The Standard Model can only be ready for the textbooks when this mechanism is finally uncovered. Higgs hunting became the primary goal of the next generation of machines.

LEP was not Europe's only trump card. LEP had been one of the two new major European projects that ECFA had been pushing at the end of the 1970s. The other was HERA (Hadron-Elektron-Ring-Anlage), a 6.3 km electron–proton collider for DESY. In some ways it was a strange combination; CERN, traditionally a proton laboratory, had to learn how to handle electrons for LEP, while DESY, traditionally an electron laboratory, had to learn how to handle protons for HERA. In HERA, the puny 30 GeV electrons do not make much impact on the incoming 850 GeV protons. HERA's electron–proton collisions are therefore a half-way house between collisions between similar particles and fixed-target experiments, where an accelerator beam is fired at a stationary target. The idea was that the electrons would penetrate deep inside the quark structure of the proton, providing a much finer proton 'x-ray' than had previously been possible. The 30 GeV electron ring, handled by DESY accelerator maestro Gustav-Adolf Voss, was ready in 1988, while the complex superconducting proton ring, with its cold-iron ISABELLE-inspired magnets and directed by Norwegian Björn Wiik, was commissioned in 1991. To monitor the collisions, two big experiments, H1 and Zeus, were built, both looking very asymmetric, with most detector equipment on the proton side.

As well as attacking new physics, HERA introduced a way of building large scientific facilities. Unlike CERN, DESY is not an international organization. Instead, it is very much a national flagship, funded mainly by the Federal German Government, but with 10% of the budget supplied by the Hamburg region. However for HERA the German Minister of Science and Technology stipulated that a share of the construction had to be borne by other nations. In the 1980s, DESY Director General Volker Soergel drew up a 'shopping list' of high-technology items needed for the new machine and looked around for partners. In return for their investment, the contributors would have the right to participate in the unique HERA research programme.

Italy supplied many of the superconducting magnets for the proton ring. The French national laboratory at Saclay furnished superconducting quadrupoles to focus the proton beams. The Netherlands supplied superconducting correction coils for fine adjustments on the proton beams. Canada supplied beam transport elements and high-frequency equipment. Poland contributed components for the vacuum system of the HERA electron ring. The Swiss national laboratory in Villigen developed the liquid-helium distribution line. Brookhaven in the USA helped with quality control of the superconducting magnets. The UK helped with radio-frequency equipment and with the injector systems. Israel's Weizmann Institute developed current carriers for the superconducting magnets. Chinese physicists and engineers provided additional help.

For a decade, major physics experiments had been international affairs, with member teams in the collaboration supplying and looking after specific components of apparatus. A laboratory such as CERN would supply the particle accelerator and associated infrastructure, but each experiment would be the subject of an international collaboration. HERA had shown how this approach could be extended to a big particle accelerator; the effort for building the accelerator did not have to be limited to the laboratory at which it was sited. A new scenario for international physics collaboration had been introduced: the 'HERA model'.

ARMAGEDDON IN TEXAS

The big accelerators in the USA attracted physicists from all over the world. Just as in the 1920s and 1930s, when it had been mandatory for an aspiring American physicist to spend a year or two at Cambridge, Copenhagen or at one of the major German-speaking universities, so in the 1950s and 1960s a new generation of scientists had to travel to gain experience. This time it was the turn of the Europeans to cut their research teeth at Brookhaven, Berkeley or one of the other US centres. In the run-up to the launch of CERN's experimental programme in the early 1960s, several budding European physicists had been expressly sent to the USA to learn their trade.

The arrival of the ISR at CERN in the early 1970s was the first time since the Second World War that an innovative particle research facility had been available outside the USA. There was an eastwards trickle as US researchers began to participate in the ISR programme. By 1978, 25% of ISR physicists were from US research centres, with Brookhaven, Columbia, Harvard, MIT, Northwestern, Riverside, Rockefeller and Syracuse groups in particular making an impact.

Detector technology was also evolving. In 1967, Georges Charpak at CERN developed the multiwire chamber, an electronic means of recording where particles had passed. As well as giving on-line information, this opened up the prospect of the 'electronic bubble chamber'—recording the tracks of particles without actually taking photographs and without complicated additional technology. Unlike bubble chambers, which record whatever comes their way, these new electronic detectors could also be 'triggered' with pre-set conditions, ensuring that only certain types of interaction were intercepted, and less interesting stuff discarded. The experiments could also be linked to computers, greatly simplifying the subsequent data analysis.

The ISR era and the advent of new detection techniques opened a door for US participation in European-based research. Several influential US figures focused their work on European laboratories; Jack Steinberger arrived at CERN in 1968 for a new experiment, where his group was the first to use the new Charpak wire chamber technique on a large scale. After an introduction to European-based research at DESY, Sam Ting from MIT was prominent in a major ISR experiment and subsequently headed the Mark-J team at DESY's PETRA electron–positron collider. Continuing his neutrino tradition, Steinberger led the first major neutrino experiment at CERN's SPS, while Carlo Rubbia's US influence brought major US participation in his historic experiment at the SPS proton–antiproton collider. With the arrival of the LEP era, Steinberger initially headed the Aleph experiment, while Ting's L3 detector, the largest of the four major LEP experiments, became a flagship for US research based in Europe. Putting all these collaborations together, by the early 1980s some 15% of US university research in particle physics was based in Europe. The irrevocable westward-ho migration of physics which had begun half a century earlier was giving way to some two-way traffic.

Fermilab, with the world's highest-energy beams, naturally continued to provide a world focus but, with Fermilab's superconducting Tevatron on a smooth course, Bob Wilson began to conjure up ideas for the next stage. From one generation of accelerators to the next had always involved a major increase in energy; going from the Brookhaven–CERN generation of proton synchrotrons to the Fermilab–CERN supersynchrotrons had increased beam energies by a factor of 15 or so. The superconducting Tevatron would double this. Wilson's imagination saw a new monster superconducting machine to reach out to 20 TeV, a startling concept soon dubbed the Superconducting Supercollider (SSC). Even though the lead time for such a giant project would be of the order of a decade, if not more, the US physics community, with no major new project flag around which to rally and with Europe breathing down their necks, was beginning to feel uneasy. Wherever and whenever the SSC was going to be built, with its circumference of some hundred kilometres it would need a large area of land. Siting the new machine was to be a story in itself, but one place in the USA where land was plentiful was the desert, and the SSC soon earned the unofficial name of 'Desertron'. Whatever it was called, the bold idea captured the spirit of the time.

Following an exploratory 20 TeV workshop at Cornell in the spring of 1982, a summer study at Snowmass, Colorado, set out to plan future directions for US high-energy physics and to explore the limits of the available technological capabilities. This meeting, attended by some 150 participants, was the spark from which enthusiasm for the new US supercollider was kindled.

For physics, the writing on the wall was clear and still is. The Standard Model reigns supreme, withstanding even the closest scrutiny by LEP but, despite its apparent impregnability, the Standard Model is not yet carved from stone. It has too many arbitrary parameters, quantities which cannot be predicted and can only be measured in experiments. The pattern of quark families and the masses of all particles remain a mystery. In the electroweak picture, the relative importance of the electromagnetic and weak sectors gives one empirical number: the 'Weinberg angle'. The relative strengths of the electroweak and gluon effects gives another. The biggest question mark of all hangs over the underlying symmetry breaking, 'the Higgs mechanism', which makes the W and

Z particles as heavy as a medium-sized nucleus, while the electromagnetic photon escapes without any mass penalty.

Theory can say little definite about the Higgs mechanism but, even with the missing top quark finally in place after its 1995 discovery at Fermilab, the parameters of the Higgs sector remain curiously impenetrable. Many people speak of a 'Higgs particle', but a single particle is a great oversimplification. The Higgs mechanism is shorthand for a whole new physics. The only confident pointer is that the equations compellingly suggest that 'something' should happen with quarks when energies reach the reciprocal of the square root of the number which governs the strength of the weak interaction: at about 1 TeV energy on the quark scale, say the theorists, something should start to show up. However, quarks cannot themselves be accelerated, only their proton envelopes. To be sure that individual quarks attain the 1 TeV threshold, the enveloping proton has to be accelerated to about 10 TeV. Although the exact form of what would happen is unclear, theorists are adamant that something new will turn up. Those swearing allegiance to the SSC billed it as a 'no-lose theorem'.

The US Department of Energy, the successor to the AEC and the principal paymaster of US particle physics, uses its High-Energy Physics Advisory Panel (HEPAP) as counsel. In 1983, a special HEPAP subpanel was set up to 'make recommendations relative to the scientific requirements and opportunities for a forefront US high energy physics programme for the next five to ten years'. Chaired by Stan Wojcicki of Stanford, the 18-strong subpanel included two Europeans, John Adams and Carlo Rubbia. After an initial organizational meeting in Washington in February, the subpanel subsequently met at Fermilab, Brookhaven and Stanford to hear presentations and, at a major meeting at the US National Academy of Sciences Study Center at Woods Hole, Massachusetts, in June, its findings were thrashed out. In addition to the presentations, Subpanel members received a mass of written material, including some 200 letters.

Specific decisions were expected on the proposed US multiteraelectronvolt high-luminosity proton–proton collider, the SSC. In addition the subpanel was also charged with making recommendations for developments across the nation: on a specific new collider proposal from Fermilab, on the future of the Colliding Beam Accelerator (CBA), ex-ISABELLE, at Brookhaven that had been dogged by disappointing superconducting magnet performance, and on future directions for technology research and development, particularly in the accelerator field.

Under the heading 'The state of high-energy physics today', the report said, 'The dominant role of US high-energy physics in the three and a half decades since the end of the Second World War has been the result of the enlightened support of the field by successive administrations and the US Congress. Its flourishing has been a source of pride among physicists, and surely also throughout the nation. It has provided a unique measure of cultural enrichment and intellectual growth for the whole world. That is not to say that others have made no important contributions. On the contrary, Western Europe, in particular, has made impressive strides in the last decade, with the PETRA storage ring at Hamburg and the SPS proton–antiproton collider at CERN, but high-energy physics is a great intellectual and idealistic endeavour that knows no national boundaries. We rejoice at other's discoveries, as they do at ours. United States high-energy physics at home and abroad builds on a strong tradition of vigour and excellence that should be maintained and nourished in the coming decades.'

The timing of the subpanel's report could not have been more significant. In January 1983, the news that CERN's proton–antiproton collider had discovered the long-awaited W particle had been acclaimed worldwide, but in the USA there had been mixed feelings. Followed up by the May 1983 announcement of the Z discovery, the effect was stunning. The W and Z discovery was one of the scientific achievements of the century. The seedling which Carlo Rubbia had offered to Fermilab but had been grabbed by CERN had now borne fruit. Coming soon after the PETRA electron–positron collider at DESY in Hamburg had given evidence for gluons, European particle physics was riding a wave.

Somewhat crestfallen, the US subpanel unanimously recommended remedial action—the immediate initiation of the SSC with the goal of physics experiments 'at the earliest possible date'. This project appeared to 'reflect the universal aspirations of the US high-energy physics community'. It provided 'the promise of important and exciting advances' in physics, while the technological challenge was considered attainable. 'This project lies within the reach of our present-day technology as a result of many years of super-conducting magnet research and development', said the report. The project 'has captured the minds of high-energy physicists everywhere,' the report went on, 'and should be initiated at the earliest possible date'. The energy goal was 10–20 TeV per beam with 'completion in the first half of the 1990s', although this was acknowledged as an 'ambitious time scale'.

The pro-SSC recommendation set the tone for the rest of the report, which found against the new Fermilab collider proposal, on the grounds that it would interfere with the SSC. Opinion on the CBA–ISABELLE scheme was divided, but again interference with the proposed SSC was cited as a major factor, and the consensus vote was not to approve the Brookhaven scheme. There were rumours that CBA–ISABELLE would have to be cancelled if the SSC were to fall on fertile ground in the new administration of Ronald Reagan. Although a continuing programme of accelerator research and development was in principle strongly supported, as far as identifiable projects were concerned the USA was putting its eggs in one SSC basket, but this package which linked

CBA–ISABELLE cancellation with SSC approval was to weigh heavily on the SSC's conscience. As Alvin Trivelpiece, Director of Energy Research in the Department of Energy, subsequently said in a 1985 interview, 'Hardly a week goes by in the Department that somebody doesn't ask if we don't have another ISABELLE situation lurking here ... There were certain identifiable failures of management and control and technology in the whole ISABELLE process that cause a greater-than-average degree of scrutiny on this, and that is likely to persist.' The SSC was to be haunted by ISABELLE's ghost.

The Woods Hole report also stressed the international nature of the field: 'Duplication of facilities with essentially the same capabilities becomes unacceptable as size, complexity and cost increase. It is, therefore, natural to contemplate US participation in outstanding ventures at foreign laboratories in order to provide greater physics opportunities for the USA and the other countries. On the other side, the USA should welcome foreign involvement in exploitation of new US facilities, in either a general way or for a specific capability.'

Following the report's recommendations, in the fall of 1983 the US Department of Energy gave the green light for preliminary research and development work for the SSC to begin, and sponsored a major Reference Designs Study, duly completed in May 1984. To carry out this work, the Department of Energy contracted with Universities Research Association, the consortium initially established to run Fermilab, to direct and coordinate the SSC development programme. However, emphasizing the nationwide involvement in the new project, work was based at the Lawrence Berkeley Laboratory, under the leadership of Maury Tigner of Cornell, who transferred to Berkeley for the duration. The work involved some 150 scientists and engineers organized into six task groups and assisted by designated laboratory coordinators in the major national laboratories Argonne, Berkeley, Brookhaven, Cornell, SLAC and, significantly, the Texas Accelerator Center, where a new group had recently been set up. The timing of the designs study aimed to provide input to a 1984 Summer Study in Snowmass, as well as providing the Department of Energy with guidelines for the major new project.

The study examined the feasibility of building a 20 TeV per beam proton–proton collider, using three different scenarios for the superconducting magnets, all based on niobium–titanium conductor cooled by liquid helium to 4.5 K. One was a high-field design, 6.5 T, with both beam tubes enclosed in a common cryostat; the second was a medium-field 5 T design with the two beam tubes in separate cryostats; the third a low-field 3 T 'superferric' design used a predominantly iron magnet. Obviously the circumferences of the designs reflected the strengths of the guiding fields: 60 km for the high-field option, 113 km for the medium-field design and a whopping 164 km for the low-field design. Even the smallest option was almost three times the size of CERN's LEP ring. Machine costs obviously varied according to the design, but hovered around the US$3 billion mark. However, just to feed each ring would need an impressive series of injectors: a linac to take the protons over the first 1 GeV, a 70 GeV synchrotron second stage, and a 1 TeV synchrotron to place the particles in orbit around the supercollider. The 1 TeV booster bore an uncanny resemblance to the Fermilab Tevatron. (The large acceleration steps in this injector chain subsequently led the design to be changed to include three booster synchrotrons—with a Low-Energy Booster of 540 m circumference   taking 600 MeV protons from the linac to 11 GeV to feed a Medium-Energy Booster of 3.96 km circumference. This would take the protons to 200 GeV, and a 10.69 km High-Energy Booster would take over to inject into the 87 km main ring at 2000 GeV.) By the middle of 1984, a mass of detailed design work had been prepared, covering all aspects of the machine. The machine would need some 20 000 km of superconducting cable, almost enough to go round the Equator, for a total of some 12 000 magnets. It was a project to fire the imagination.

With ideas for the big machine beginning to take shape, it was time to start to plan for the experiments, the big eyes to watch the supercollisions. In July 1984 a summer study on the 'Design and Utilization of the Supercollider' looked at the challenges of research at this new frontier. To ensure capturing enough rare events, the supercollider would have to reach for a very high collision rate (related to a quantity called 'luminosity' in the trade). As well as threatening to swamp detectors in tidal waves of data, such an intense particle bombardment might also harm the detectors supposed to measure it.

After building test magnets of various designs, in August 1985 the SSC Magnet Selection Advisory Panel recommended a compromise ('Design D') magnet which combined the best features of the high-field (which had been explored at Berkeley and Brookhaven) and medium-field (explored at Fermilab) designs, but opting for separate cryostats for the two proton lines. All Design D test magnets reached a field strength of 6.6 T or more, promising well for the continued development work towards a reappraised design field for the final machine of 6.6 T.

In 1965, it had been the US National Academy of Sciences and the US National Academy of Engineering which had chosen the site for the 200 GeV proton synchrotron project eventually to become Fermilab. Twenty years later, this process was repeated when Alvin Trivelpiece enlisted aid from the US National Academy of Sciences and the US National Academy of Engineering in the selection of an SSC site. The selection would be handled by a review panel of 15 distinguished experts. As well as the tunnel itself, buried some 12 m below ground, the SSC would need a cluster of buildings every 8 km or so to house cryogenics, power supplies and other support installations. At several points in the ring, immense underground caverns would be excavated, 50 m wide and long, to house the physics experiments. The focus of the site would be a new campus with buildings and facilities to house a staff of 3000.

The formidable intellectual power of the US physics community was mobilized to push the SSC. Books, talks, brochures, press articles and television appearances were all influential, while lobbying in Washington was a new skill that had to be acquired. In November 1985, a meeting of 57 senior physicists at Berkeley issued a statement: 'As a group with varied responsibilities within the US high-energy physics community, we wish to emphasize the vital importance of the SSC for the future of elementary particle physics. An accelerator probing the several TeV mass scale by the mid-1990s is indispensable for continuing the remarkable recent progress in understanding the fundamental structure of matter. ... However, to maintain this date as a realistic target, increased SSC research and development funding is a necessity. ... The SSC is essential for exploring the new energy region and will reveal fundamental aspects of the new synthesis.'

Having done their duty and supported the new US national flagship, the existing national laboratories had to ensure that their budgets were not raided to provide new funds for the SSC. At the same time as stressing the need for continued research and development work on the SSC scale, in 1985 HEPAP drew attention to a range of existing US commitments. Research was moving forwards at Fermilab, Cornell and Stanford, and this effort, together with the large US investment in the L3 experiment at CERN's LEP electron–positron collider, had to be safeguarded. 'Awaiting the arrival of the SSC, the particle physics front-line should not waver,' urged HEPAP.

In March 1986, the SSC 'Conceptual Design Report', 712 pages of it, was submitted to the Department of Energy. With its bottom line cost estimate of US$3 billion for the machine and a 6.5 year construction schedule, the conclusion was that 'the design is technically feasible and properly scoped to meet the requirements of the US high-energy physics community from the mid-1990s to well into the next century.' It also stated that 'the SSC cost estimate is credible and consistent with the scope of the project', and 'the proposed 6.5 year construction schedule appears feasible for the assumed funding profile and for the reasonable assumptions made concerning the characteristics of the site'.

On the design front, the SSC had evolved from a circular into an 84 km oval racetrack shape, with the collision areas sited near the straighter sections of the ring to facilitate catching particles which continue to follow the direction of the colliding beams. On 30 January 1987, US physicists were jubilant when President Ronald Reagan gave official approval for construction of the SSC. With the requested start in 1988, the project could be complete by 1996 if sufficient funding were assured.

This was followed on 10 February by the announcement of a schedule for submission of proposals for candidate sites. 43 proposals were received from 25 states, from Florida to Alaska, and from New York to Washington, together giving some 300 kg of documentation. The 43 proposals were soon whittled down to 36, after which selection became more difficult. As well as geology and suitability for tunnelling, criteria included regional resources such as housing, education, culture and environment, as well as cost. The short list of best-qualified sites emerging from the arduous selection were in Arizona, Colorado, Illinois (with a site adjoining Fermilab), Michigan, New York, North Carolina, Tennessee and Texas.

In parallel, plans for the big detectors began to take shape, with attention focusing initially on large general-purpose detectors to catch as many of possible of the particles emerging from the supercollisions. These would be huge affairs, four stories high, weighing 40 000 tonnes. Despite the enormous size of such detectors, they would be packed with smaller-scale equipment, for

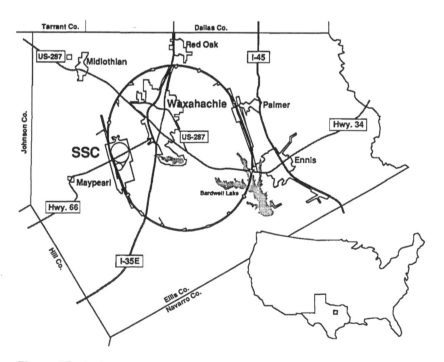

Figure 47: *As big as a city—the proposed site of the 87 km SSC ring in Ellis County, south of Dallas, Texas.*

which research and development was already under way at universities and laboratories all over the world.

In 1989, progress shifted gear when the site for the SSC was chosen in Ellis County, south of Dallas, Texas. The Lone Star State, hungry for culture, had prevailed, accompanied by the promise of a US$1 billion of state money to help construction. The Director of the new SSC laboratory was Roy Schwitters of Harvard, formerly co-spokesman of one of the major experiments at Fermilab's Tevatron collider and a key member of the team which had discovered the J/psi at Stanford's SPEAR ring in 1974. SSC staff moving into temporary accommodation near Dallas were heartened by the news of US$225 million of federal money. The following year this was increased by 40%.

The figure of US$3 billion for construction of the machine cited in the 1986 Conceptual Design Report had deliberately not taken account of preliminary site preparation. It also omitted the mighty detectors needed to exploit the huge machine. Adding in these factors, together with the extra year of construction time now foreseen, gave an initial baseline cost of US$5.9 billion for building the SSC in Texas.

However, clouds were gathering on the horizon. When George Bush took over as President in 1989, Admiral James Watkins was nominated as Energy Secretary, replacing John Herrington, who had been a staunch SSC advocate. Watkins ran his Department in military style, preferring construction engineers to accelerator specialists in key project positions. On site, Joseph Cipriano reported directly to Washington, bypassing senior officials as well as Schwitters. This created a large local staff whose roles and responsibilities overlapped with those of SSC personnel. The traditional gung-ho approach to building large US accelerators had been compromised.

On the cost side, the original figure of US$5.9 billion had escalated towards the US$8 billion mark as new design decisions were taken and more margin was built in. With the planned construction time continually being stretched, the cost figure showing no sign of stabilizing and a worried Congress committee proposed a ceiling of US$5.9 billion from US funds, a figure which assumed that some US$2 billion of additional SSC aid would come in from overseas. However, such a ceiling would constrain the SSC to a proportionally smaller ring, providing a beam energy of 15 TeV. A specially convened group of the HEPAP concluded, 'The SSC laboratory, in its current proposal has chosen wisely among the alternatives and that the project should move forward expediously. ... Lowering the beam energy to 15 TeV would unacceptably increase the risk of missing important new physics. We feel strongly about the need for a flexible and reliable facility at 20 TeV for decades to come, and therefore we believe it would be very unwise to redesign the machine with a reduced circumference.' In Congress, there were anxious calls for foreign contributions to boost the SSC effort.

With the first magnets arriving on the Texas site from Brookhaven and Fermilab for testing and examination, the actual 'footprint' of the machine was decided, with the ring surrounding the town of Waxahachie. The first priority was to prepare for construction on the 'green field' site, and first land was purchased. A total of 16 553 acres would have to be acquired from 707 different landowners. A shaft 16 ft in diameter, and 250 ft deep to explore subterranean Texas was soon joined by a major 60 ft × 30 ft shaft, big enough for equipment to be lowered to the tunnel below. Awaiting the completion of the on-site magnet development laboratory, itself as big as a football pitch, 17 m superconducting magnets were tested at Fermilab.

As more physicists organized themselves into groups to design and build the modules for SSC detectors, the detector designs took shape. In response to a formal SSC invitation, letters of intent for major experiments began to arrive. A

Figure 48: *1990: drilling began on the SSC site.*

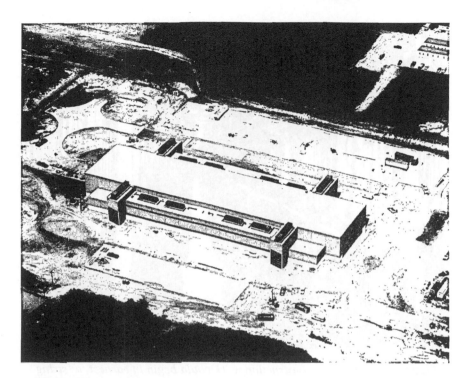

Figure 49: *One of the football-pitch-size buildings erected on the SSC site. (Photo SSC Laboratory.)*

major contender was the Solenoidal Detector Collaboration (SDC), later joined by (Gamma, Electron and Muon (GEM)). Meanwhile the 1991 Gulf War successes boosted US national pride and gave additional impetus to the SSC effort.

Before construction work could begin in earnest, a 'string test' of five large superconducting magnets assembled on the Texas site in a building 220 m long had to show its mettle. This milestone had been set for October 1992 but was in fact passed in August—a real morale boost to SSC staff. With the construction light now switched to green, a boring machine was lowered to the bottom of the big shaft and in early 1993 set out on the first few kilometres of tunnel. Construction also began for the linear accelerator to provide the SSC's protons with their first step up in energy. Off site, industrial suppliers General Dynamics and Westinghouse prepared major plants for large-scale magnet production. Following the 'HERA model' approach pioneered at DESY, discussions were under way with research centres overseas who had showed an interest in providing equipment for the new machine. China, India and Russia had indicated their willingness to provide equipment.

Figure 50: *Before SSC construction work could begin in earnest, a 'string test' of five large superconducting magnets had to show its mettle. This milestone had been set for October 1992 but was in fact passed in August—a real morale boost for the SSC.*

Then on 17 June 1992 came a stab in the back. The US House of Representatives, preoccupied with balancing the US budget, singled out the SSC as a scapegoat. The decision to site the SSC in Texas had rankled several states and increased the SSC's visibility in Washington, but a vigorous pro-SSC faction had always come out on top. This time Joe Barton of Texas, a prominent pro-SSC member of the House of Representatives, was caught off-balance by the surprise motion. 'The mood of the Congress, or of the House at least, was to take a scalp and take it home,' he said later. In a 232–181 vote, a carefully-constructed package of energy-related items excluded any significant ongoing SSC funding. The President's budget had originally asked for US$650 million of funding for 1993, which immediately had been trimmed by the House to US$484 million. After the vote, only US$34 million of closedown money was left on the table. 'It's an enormous pig in a poke,' said Howard Wolpe of Michigan. 'We can't continue spending money on projects we don't need,' echoed Dennis Eckart of Ohio. In the House of Representatives, the Republican Texas delegation had been prominent in trying to push through a

Balanced Budget Amendment to curtail US public overspending. The motion did not survive, but Democrats had been angered by the Texas delegation's efforts and were looking to retaliate. The SSC vote showed how tough they could be.

On 30 July, President George Bush visited the SSC, promising 'to fight hard and continue to fight hard' for the machine. On 3 August, a pro-SSC motion in the Senate was voted in, 62–32. It was a rerun of the 1990 Space Station affair, which had been voted down in the House but subsequently reinstated by the Senate. The SSC had volunteered to host the big international physics meeting for 1992 in Dallas, and the atmosphere of relief there was almost tangible. During the weekend break, the thousand delegates were bussed out to the SSC laboratory to admire superconducting magnets under test, to peer down the installation shaft, and to enjoy a Texas barbeque, with strict instructions not to touch any poison ivy.

To safeguard the SSC's progress in subsequent Congress votes, further promises of support from other nations, notably Japan, were going to be vital. Overseas money had already been built into the cost estimates, but the cash was not yet on the table. However, the US presidential election campaign and the arrival of Bill Clinton in the White House temporarily blinkered US vision of international affairs. The negative effect of the 1992 House of Representatives vote also did not encourage other nations, particularly Japan, which had been pondering about its support for the SSC. In December 1992, an intergovernmental agreement on SSC collaboration was signed in Moscow by the US Department of Energy and the Ministry of Atomic Energy of the Russian Federation. Russian centres would provide SSC equipment. On site in Texas, work progressed, with 12 shafts having been sunk to the tunnel depth. With four tunnel boring machines at work, over 6 km of tunnel had been bored by April 1993.

However, behind the scenes, the opposing parties began limbering up for the next round of the Congress voting match. The Administration planned to request Congress to appropriate US$640 million for 1994 funding, a substantial increase over the US$517 million finally voted for 1993. However, funding erosion and price increases meant that the SSC's completion date had meanwhile slipped to 1999. In the House of Representatives, sniping had already started. The Chairman of the House Committee on Science, Space and Technology, George Brown of California, released a report prepared by the General Accounting Office, saying, 'The GAO has written a report claiming that the SSC is over budget and behind schedule. In response, the Department of Energy contends that the GAO has conducted a faulty analysis based on insufficient data. ... Our goal in the coming weeks will be to understand ... these differing views.'

Hazel O'Leary, Energy Secretary under the Clinton administration, mounted a staunch defence, claiming that the project was being managed 'in a very conservative manner', and had used only 3.5% of its contingency. Milestones were being met, and in particular the crucial superconducting magnet string test had been successful six weeks before its appointed deadline. Sensitive to such attacks and allegations, the Department of Energy had installed more federal staff at the SSC laboratory to monitor progress. This was widely resented by SSC personnel, who saw it as questioning their competence and undermining their authority. O'Leary said, 'The SSC laboratory has lacked the leadership and consistent good judgment which I believe are essential for success of this project. ... This project does not have the sort of team relationship and project leadership that exists in other Departmental construction projects.' SSC Director Schwitters retorted, 'The SSC is perhaps the most reviewed and overviewed construction project in history.'

With the SSC under attack from several quarters, the pattern for 1993 was emerging. In Texas the SSC itself pushed ahead vigorously; the world admired fresh superconducting magnet achievements and tunnelling progress, while plans for the big experiments continued to take shape and attract new collaborators. However, an organized group of politically powerful SSC baiters had got their act together. They set out to erode annual funding to distort the accounting while demanding that baseline estimates in Texas should still be met. They would decry the lack of foreign interest while at the same time undermining international confidence in the project.

More visible was the shadow being cast by the ballooning US trade deficit. Prime-time television slots drew attention to Washington overspending, highlighted by facile case histories about allegedly lavish parties and catering. A typical item was the 'discovery' of some US$260 000 earmarked for 'discretionary spending' by SSC management. The SSC maintained that more than half of this sum was to pay for meetings, expenses and hospitality for the many experts from around the world who visited the laboratory regularly, without pay, to help the project. In a bid to make the SSC look more attractive in such an adverse climate of opinion, the basic science message was camouflaged by secondary arguments—the SSC would create jobs, improve the nation's acknowledged poor mathematical literacy and provide particle beams for cancer treatment.

With the anti-SSC lobby becoming so vocal, it was not surprising that an anti-SSC motion, packaged as an amendment to the normal appropriations bill, was passed in the House, this time by the more significant margin of 280 votes to 150. In 1993, the House had many new representatives, many of whom were keen to be identified with politically correct cost-cutting moves. In this motion, SSC funding was reduced by US$400 million, leaving only US$220 million to wind up the project in an orderly way. While the SSC and its supporters hoped for a repeat of the previous year, with final victory in the Senate, the outlook was less optimistic. Texas, a traditional SSC supporter, deferred its own US$79 million funding for the remainder of 1993. With this income already allowed for in SSC spending, the brakes had to be slammed on. To prepare the way for the Senate discussions, a major public meeting was staged in Washington. Petitions and letters of support flowed in.

Again the Senate looked as if it would be able to reverse the earlier House vote. Led by SSC champion Bennett Johnson, the upper house again turned the tables. There was a sigh of relief from the SSC lines. When the two houses disagree, the contentious bill becomes the subject of arbitration or 'conference' to resolve the disagreement. On 14 October, an initial conference report recommended giving the SSC everything it had asked for, but the SSC opponents were not finished this time. Incensed by what they saw as biased conferencing, the bill was rejected and went back to conference, emerging this time with an opposite recommendation. This was approved by the House by 332 votes to 81 and endorsed 89–11 by the Senate. On 28 October 1993, President Bill Clinton signed the SSC's death warrant. A one-time offer of US$640 million was made available to wind the project down. Roy Schwitters resigned, and 23 km of underground tunnel in Texas were to remain empty. Fermilab Director John Peoples had the unenviable task of dismantling the SSC infrastructure.

In the heady days of the 1980s, when deficit spending could be underwritten by 'junk bonds', an ambitious project such as the SSC had captured the public imagination but, in the cost-conscious 1990s, what had gone up like a balloon went down like a weight. The awe-inspiring project which was to have restored US scientific supremacy had been assassinated by legislation. Initially hailed as the scientific parallel of the Strategic Defense Initiative ('Star Wars') project, it crept out through the back door as 'The Vietnam of particle physics', as one SSC victim put it. It took too long to get going, and, with its ballooning price tag and Texas connections increasingly visible from afar, fell victim to a change in the national political and economic climate. It suffered a similar fate to its companion project, Star Wars, which had foundered after the abrupt reversal of the international political tide with the disappearance of the Cold War. However, premature cancellation did not necessarily imply that Star Wars had been unsuccessful. One goal had been to intimidate the Soviet nuclear arsenal, and this impressive harnessing of US scientific and technological power surely affected Soviet sentiment and contributed to the sudden appearance of perestroika.

With the abrupt decision to cancel the SSC, particle physics careers were in ruins, and an estimated 2000 highly qualified people were looking for a job. With their sophisticated computer skills, many were siphoned off as mathematical modellers by astute fund managers trying to keep one jump ahead of the financial markets. Half-way around the world, magnets for an SSC injector synchrotron manufactured at the Novosibirsk laboratory in Russian Siberia would remain forlornly stacked. If CERN's foreclosure of its ISR had been 'scientific vandalism', the cancellation of the US SSC was deliberate assassination. On the scientific front, the aims of the SSC beckoned as strongly as ever, while progress on the technical side had been cautious but sure. The assassins had been politically motivated.

Ironically, the CBA (formerly ISABELLE) machine, which had been given a death sentence by the initial recommendation for the SSC, was disinterred from its Brookhaven grave. An alternative option, handling beams of heavy nuclei, had been prepared, and in August 1994 the first magnet was installed in the ex-ISABELLE tunnel for the Relativistic Heavy-Ion Collider (RHIC) fed with nuclei by the Brookhaven AGS and scheduled to begin operations in 1999. Also, on 4 October 1993, President Clinton announced the decision to build a new 'B factory' at SLAC. The US$36 million of initial funding for this electron–positron collider, built in the same tunnel as the PEP ring, had been rejected in the SSC package voted down by the House. Unlike the SSC, this managed to survive the subsequent Senate–House negotiations. The resourceful Fermilab, whose plans had also suffered a major setback at the time of the initial SSC recommendation, tabled a plan for a 3.6 km Main Injector to replace Bob Wilson's original Main Ring and to feed the mighty Tevatron. SLAC continued to work towards a higher-energy electron–positron collider. Cornell's CESR electron–positron collider was upgraded. A new electron machine, the Continuous Electron Beam Accelerator Facility (CEBAF), later renamed the Jefferson Laboratory, was built at Newport News, Virginia. Like RHIC at Brookhaven, this was designed to cater for nuclear as well as subnuclear physicists. The USA still had an impressive research flotilla, but now it lacked a flagship.

LAST IN THE LINE OF RINGS

In 1976, with synchrotrons becoming progressively larger, the Commission for Particles and Fields of IUPAP set up a study group to investigate the possibility of a Very Big Accelerator (VBA). However, this was not another machine design group to produce a blueprint for a supermachine. Instead the idea was to explore avenues for increased international collaboration; the VBA would be a dream machine, an interregional project bringing together the traditional geographical research concentrations in basic physics: the USA, Europe, the Soviet Union and the rest of the world. This led to the establishment of the International Committee for Further Accelerators (ICFA), as an interregional forum. ICFA's first elected chairman was Bernard Gregory, who unfortunately died at Christmas 1977 before ICFA's business had got into its stride. His place was taken by John Adams, not the first time that Adams had stepped into a dead man's shoes. Although its attention was initially focused on the specific VBA goal, ICFA's sights slowly shifted towards interregional collaboration in general. International collaboration in accelerators is what ICFA is all about and, with the US SSC looking to interest foreign participants, ICFA would have been a good forum to launch such ideas. However, the US venture chose to bypass ICFA.

International awareness of high energy physics was growing. The subject was identified as a topic ripe for discussion at specialist subgroups in the framework of the 'G7' summit meetings of the world's major industrial powers. The influential Organization for Economic Cooperation and Development (OECD) in 1992 established a 'Megascience Forum' in which major international scientific ventures were discussed and in which particle physics projects featured prominently. With OECD membership restricted to financially powerful nations, UNESCO, CERN's foster-parent, created a Physics Action Council which promoted physics interest in developing countries. Although these political platforms provided useful dialogue and continue to do so, they followed, rather than influenced, the progress of major physics projects. At a political and diplomatic level, the scientific message can be diluted or distorted. Plans for new big machines evolved from the bottom up and not from the top down. Nevertheless major international projects need powerful political lifting gear to move heavy pieces into position.

As moves for the SSC gathered momentum on one side of the Atlantic, in March 1984, initially at Lausanne and subsequently at CERN, physicists met for a workshop 'On the feasibility of a hadron collider in the LEP tunnel'. At the

Figure 51: *French President François Mitterrand speaks at the official inaguration of CERN's LEP machine on 13 November 1989.*

time, large-scale construction, under LEP project leader Emilio Picasso, was getting under way for CERN's new LEP electron–positron collider, with shafts being sunk for the 27 km tunnel. LEP groundbreaking had been marked by an official ceremony on 13 September 1983, with French President François Mitterrand as guest of honour.

The 27 km LEP tunnel was a major investment from which maximum returns had to be ensured. Almost from the outset, the tunnel's dimensions had been made large enough to leave room for another machine: a subsequent ring of bulky superconducting magnets to handle a high-energy proton beam, in what John Adams had initially seen in 1977 as a mighty European proton–electron complex. This proton synchrotron was soon obscurely christened the Large Hadron Collider (LHC) (hadrons are subnuclear particles which contain quarks and which interact through the colour force). Because of the problem of contending with different languages and cultures, CERN's machines had traditionally been given obscure dull acronyms. There were no Cosmotrons, Bevatrons, Tevatrons or Supercolliders in Geneva.

With the LHC scheme still under wraps, other ideas for big new machines were being pushed. With LEP having introduced CERN to electrons, ideas were turning towards the possibility of developing a big linear accelerator to fire electrons and positrons at each other, SLAC style. In 1985, CERN Council asked Carlo Rubbia to chair a special Long-Range Planning Committee to evaluate these various future routes.

The electron–positron physics route is clearly signposted, with electrons and positrons (point-like particles with no known constituents) annihilating. The supplied energy is converted into new particles, providing very 'clean' physics conditions uncluttered by spectator debris, but the technology needed to build such linear colliders (see page 155) was out of reach at the time, and a major programme of research and development work was called for even to arrive at a preliminary design. On the one hand, all the accelerator expertise needed to build the LHC was already in place. Admittedly the powerful superconducting magnets required were a technological challenge, but the problems were already in the engineers' court. On the other hand the LHC's physics would be 'dirty'; individual quark and gluon interactions deep inside the colliding proton envelopes would have to be identified through a dense fog of debris.

The Rubbia Committee faced a dilemma: should CERN go for an experimentally 'dirty' proton–proton collider which was not too far from existing technology or should it opt for a 'clean' electron–positron collider which was unexplored territory as far as accelerator expertise was concerned? The LHC was a big technological bite, but superconducting magnet development at Fermilab and at DESY had shown how cryogenics could be put to work for physics. The LHC bite was acknowledged as being big, but nevertheless chewable. The LHC also continued CERN's tradition of interlinking its machines. Ever since the decision had been taken to build the SPS alongside its existing machine, CERN's trump card had been its interlaced beam transport system, with one synchrotron ring feeding the next.

The recommendation that emerged from the Rubbia committee was for the LHC as CERN's primary objective, while work for a longer-term electron–positron linear collider should continue in the background. To oversee this plan, in 1989 Carlo Rubbia moved into the driving seat as CERN's Director General, succeeding Herwig Schopper, who had steered LEP from drawing board almost to reality. Director Generals with fixed-term mandates bequeath the fruits of their labours to their successors. In his first year of office, Rubbia presided over LEP's commissioning.

LEP's designers had carefully positioned the electron and positron beams and ancillary services so as to leave enough room above the ring for the LHC to be installed later. Unlike previous iterations with the SPS and with LEP, where new injectors had been proposed, from the outset it was clear that injection for the proton ring in the LEP tunnel would be via the existing machines. In this way, the LHC would become an extension of CERN's particle beam infrastructure, opening up additional possibilities, colliding other types of beam: protons, antiprotons, electrons (from LEP), or whatever else CERN had to offer. In 1986, CERN had begun a new programme of research using beams of nuclei in the SPS. These too could be fed to the LHC, providing an extended menu of LHC beams, in contrast with the SSC's fixed diet of protons.

In the USA, design work for the SSC was proceeding apace; a detailed reference design had emerged from the study group centred at Berkeley, while superconducting magnet research and development work was under way at major US laboratories. For a giant new proton collider for the turn of the century, it looked as though the SSC had a head start. However, CERN had all the infrastructure and was already starting to build its tunnel, albeit only about a third of the size of that envisaged for the SSC.

To achieve the stipulated target of 1 TeV collision energy at the quark level, to be on the safe side the LHC design initially aimed for 10 TeV beams. Bending such high-energy beams round the 'tight' confines of the 27 km LEP tunnel called for powerful magnets, providing fields of about 10 T. This was in sharp contrast with the US SSC where, although the beam energy was higher (20 TeV as against the LHC's initial target of 10 TeV), the ring was much bigger, 87 km against 27 km, so that the SSC needed a magnetic field of 'only' 6.4 T. To power such superconducting magnets, the LHC had been committed from the outset to superconducting technology. With diminished electric resistance at liquid-helium temperature, the large currents needed to generate such high fields can be forced through narrow conductors, making the magnets more compact. With standard superconducting magnets providing fields in the range from 4 T (for Fermilab's Tevatron) to 6 T (for the proton ring of DESY's HERA electron–proton collider) or 6.6 T for the SSC, attaining such high fields uniformly around such a large circumference was not too far from what had already been achieved but nevertheless was the major challenge which would have to be met if the LHC were to become a reality. Work began on the building of prototype LHC magnets but, with the LHC not yet an official project and with money earmarked for LEP, the initial level of LHC magnet activity was modest.

The next question was the choice of particles for the LHC. CERN's ISR had been built as a proton–proton collider but landed up colliding other beams, deuterons and alpha particles, as well as operating as a proton–antiproton collider. The ISR had been launched when there was no prospect of a proton–antiproton collider on the horizon. The advent of beam-cooling techniques opened the door to a proton–antiproton collider at the SPS and, with this project having been so successful, it was natural that the first ideas for the LHC foresaw a proton–antiproton collider using a single ring, a giant version of the SPS collider. However, the physics goals of the SPS collider had been clear-cut and were within the grasp of a relatively modest collision rate. When the W and Z particles were discovered in 1983, the collision rate of protons and antiprotons in the SPS was only about 0.2% of the proton–proton collision rate achieved at the ISR, but the ambitious physics aims of the LHC called for a collision rate considerably higher even than that achieved by the ISR. Attaining such a high collision rate with antiprotons would be difficult. Any antiproton

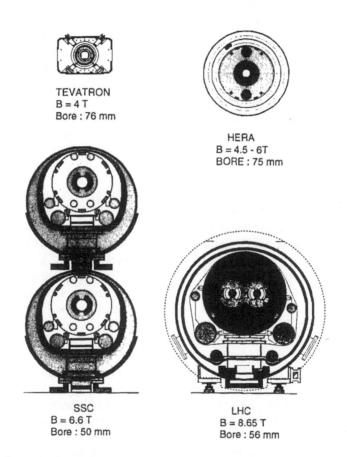

TEVATRON
B = 4 T
Bore : 76 mm

HERA
B = 4.5 - 6T
BORE : 75 mm

SSC
B = 6.6 T
Bore : 50 mm

LHC
B = 8.65 T
Bore : 56 mm

Figure 52: *Different superconducting dipole magnet fashions. Fermilab's Tevatron handles counter-rotating protons and antiprotons in a single beam channel. The room-temperature iron magnet yoke is outside the cryostat. DESY's HERA, this time with cold iron, handles protons in a single ring, the electrons having their own ring. The SSC would have used separate cryostats for its two proton channels while, to save space, CERN's LHC uses a single cryostat for its twin proton channels.*

collider system has to live with the handicap of having to invest 300 000 protons to produce a single antiproton.

While surpassing in a 27 km ring the collision rates that had been achieved in the 1 km ISR ring would still be a considerable achievement, it was clear that the proton–proton option was the only viable route. With two separate proton rings intersecting only at the collision points, the interference between the circulating beams is minimized. However, this brought the LHC design up against its first major design decision. With the 3.8 m diameter of the LEP tunnel already fixed, there was hardly room to accommodate two cryogenic systems to handle separate proton beam channels. The two superconducting rings carrying the LHC proton beams would have to be fitted into a single cryostat. With the LHC high-field superconducting magnets already a tough assignment, such a complex design would bring additional problems. Locked together in a shared cryostat, the magnets for the two rings cannot be independently aligned, and tiny field errors cannot be matched. Small correction magnets would have to be installed all around the ring to fine-tune the beam but, as well as saving space, the 'two-in-one' solution would also call for less material and would optimize cryogenics, thereby saving money.

Heading the initial LHC magnet effort was Giorgio Brianti, an Italian engineer who had joined CERN almost at the outset in 1954, when he immediately became involved in industrial liaison work for the production of equipment, particularly magnets, for CERN's first machines. Brianti went on to occupy key roles in subsequent major projects, including supervision of the conversion of the SPS from a classic proton synchrotron to its new role of proton–antiproton collider. In 1981 he became CERN's Technical Director and pushed the new programme of development work for LHC superconducting magnets, which required close collaboration with industrial specialists. After forty years, Brianti retired from CERN in 1995, but his stamp is all over the LHC magnet design.

With the choice of a two-in-one cryostat made, the next task was to decide on the superconducting material. Conventional superconducting cable is made of a niobium–titanium alloy, operating at liquid-helium temperature (around 4 K, i.e. 4 degrees above absolute zero). Still higher performance is provided by a niobium–tin alloy and, in 1989, an LHC niobium–tin test coil attained 10.2 T at 4.3 K. To match this performance with conventional niobium–titanium alloy would require boosted superconductivity at an even lower temperature, around 2 K, calling for a more complex and powerful refrigeration system, but the potential cost savings of operating a 27 km ring at 4 K rather than 2 K were obviated by the fact that niobium–tin alloy is an extremely difficult alloy to work with. As it is very brittle, it cannot be wound into coils, the only solution being first to wind the niobium and tin components, and then to heat the coil to 700 °C to produce the superconducting alloy. In addition, the coils would have to be electrically insulated, and the insulation would have to be capable of withstanding this inhospitable heat treatment. In any case, not enough of the material was available.

Superconductivity is fragile, limited by both a critical field and a critical current, above which superconductivity is destroyed and the sample reverts to normal electrical conduction; the superconductor is said to 'quench'. Helium liquefies at 4.2 K, a temperature where suitable materials become superconducting, making liquid helium the medium of choice for superconducting applications. Producing 10 T fields with niobium–titanium cable needed temperatures below 2 K. Liquid helium never solidifies at atmospheric pressure, no matter how much it is cooled. However when cooled below 2.2 K, it undergoes another quantum structure transition, becoming a new variety of liquid helium with remarkable properties. This 'superfluid', with almost zero viscosity, can flow through smaller holes than any other liquid, draining away through micropores even a millionth of a centimetre across! Although superfluid helium shows zero viscosity in this way, it still needs force to drag a large object through the liquid—the so-called 'viscosity paradox'. Placed in a container, superfluid helium forms a thin film on the walls and subsequently flows through these films, so that the superfluid defies gravity and flows out of any open container. In another typical example of scientific understatement, physicists acknowledge these remarkable properties of superfluid helium by calling it 'helium II'.

At superfluid-helium temperatures, the superconducting critical current and field move higher, allowing the magnets to carry more current while still remaining superconducting. Superfluid helium has incredible thermal properties: it conducts heat ten thousand times better than copper, while its viscosity, lower than that of ordinary liquid helium, means that cooling becomes more efficient. However, operating at a lower temperature increases both costs and the risks to the cryogenic system, where stray heat leaks can cause quenches. Superfluid helium enables kilowatts of refrigeration power to be carried over more than a kilometre with a temperature variation of less than 0.1 K. Superfluid-helium technology had been pioneered in Europe at the Grenoble centre of the Commissariat à l'Energie Atomique and exploited in the Tore Supra tokamak experiment on controlled thermonuclear fusion. Niobium–titanium alloy operating at 1.8 K soon became the preferred LHC route, and model magnets were regularly attaining fields of more than 9 T.

In another example of the technical mastery needed for twenty-first-century science, ultra-precision plumbing is needed to hold a superfluid in the 27 km LHC ring. Cooling and operating this cryogenic ring would require a powerful refrigeration system, the world's largest operating at superfluid-helium temperatures. Installing from scratch the refrigeration system for such a scheme would be a daunting task. However, here another example of interlocking facilities worked in CERN's favour, this time for cryogenics rather than particle beams.

To accelerate its electrons and positrons, LEP had been fitted with 128 radio-frequency cavities, fed by klystrons, to pump energy into the circulating beams. This complement of cavities would supply enough energy to take LEP's particles to some 50 GeV per beam and to manufacture Z particles but, after LEP's initial precision Z study, the aim was for the machine's energy to be boosted, almost doubled, to enable LEP to produce W particles. LEP was initially tuned to the electrically neutral Z, produced singly in electron–positron annihilations at a collision energy of 91 GeV (45.5 GeV per beam). However, the electrically charged W particles can only be produced in these annihilations as oppositely charged pairs, requiring a collision energy of 160 GeV (80 GeV per beam). Such a boost in energy would need a considerably increased complement of radio-frequency cavities to provide the additional power. Just as superconducting magnets can attain higher fields more economically than conventional magnets, so superconducting cavities can (depending on the operating frequency) attain higher electric fields for the same input power. The 200 or so LEP superconducting cavities, made of copper with a special coating of niobium, have to be manufactured with extreme care and prepared in ultra-clean conditions to avoid microscopic impurities and defects which would otherwise quickly mar the superconducting performance. As well as the cavities themselves, the radio-frequency feeds have to withstand extreme conditions, and perfecting the quality control on all these precision components delayed higher LEP energies by about a year.

To run these superconducting cavities, the LEP energy upgrade had called for a liquid helium supply system based on four 12 kW refrigeration plants, together about 50% larger than the cryogenics built to service the 6.3 km superconducting proton ring of DESY's HERA proton–electron collider. The new plant began regular operation for LEP in 1995, but from the outset the plan was for it to be used eventually for the LHC cryogenics, suitably adapted to run at 1.8 K rather than 4 K as in LEP.

Comprehensive LHC design studies were published in May 1991 and a revised version in November 1993. With many ideas still in the melting pot, these design studies were not binding but served as useful references to guide ongoing work. In 1994, a radically new design idea emerged. While the whole idea was that LEP and the LHC would have to cohabit the same tunnel, LEP had got there first, and it was natural that LEP would get the best seat—first come, first served. Thus the initial designs placed LEP close to the tunnel floor, leaving enough space above for the LHC to squeeze in later. Bulky LEP radio-frequency equipment had been mounted crossways so that it would not subsequently impede LHC installation above.

The LHC construction schedule published in 1991 foresaw large-scale magnet production beginning in 1993, with installation of LHC components commencing in 1995. With such installation work in the underground tunnel clearly incompatible with LEP operation, the schedule allowed for alternate periods of LHC installation and LEP running from 1993 to 1996, followed by slightly over a year of final LHC installation, during which time LEP would not run. The LHC would be commissioned in 1998, after which both machines could initially run in parallel.

With the SSC and the LHC facing each other across a scientific poker table and playing for high stakes, neither party wanted to blink first. However, when the SSC was cancelled in 1993, the poker game was off, and a more relaxed schedule could be drawn up. Stretching the LHC construction period to cover the first few years of the next millenium meant that LEP's scientific programme would be largely complete by the time that the LHC arrived. Rather than perch the LHC, with its heavy magnets and cryogenics, above an old LEP ring, it would make more sense to remove the relatively light LEP equipment so that the heavy LHC components could rest on the floor. As well as simplifying the installation, this meant that the LHC ring no longer had to follow exactly the same path in the tunnel as LEP. An optimal LHC geometry simplified the magnet configuration. This radical new design led to a new specification, published in October 1995.

However a big attraction in having both the LHC and the LEP in the same tunnel was that electrons from LEP and protons from the LHC could be collided, giving a super-HERA for almost nothing. To preserve this attractive option, enough space is being left above the LHC for subsequent reinstallation of an LEP electron ring, probably in a totally new disguise.

Before the October 1993 decision to cancel the SSC, the two giant projects, the SSC and the LHC had been pitted against each other in a race to discover the mysterious Higgs mechanism. Work for the SSC had begun earlier, with superconducting magnet development work at a more advanced stage, but the LHC's trump card was that its tunnel was already there, although its smaller circumference meant that the LHC beams needed more steering than those of the SSC; the LHC magnets were handicapped because they needed to be very powerful superconducting magnets. Making demonstration superconducting magnets to attain the required field strengths did not appear to be a problem, but filling a 27 km ring with such magnets and keeping them going day and night for weeks on end would be another matter. As prototype LHC magnets became available, they were linked together in a 'string' and run day and night to simulate actual LHC operation and to improve running efficiency.

The SSC's trump card had been its higher energy, 20 TeV as against the LHC's initial target of 10 TeV. With physics requiring 1 TeV at the quark level, the LHC was clearly at a disadvantage, but the total proton energy is shared among its quarks in a very complicated way. Extrapolating their existing knowledge of the proton, physicists knew how the component quarks would share each proton's total energy at 10 TeV and, although the LHC's protons had a lower overall energy, at the end of its quark energy wedge there would still be as much energy as those quarks that the SSC would have provided. However, these high-energy quarks would be rare. To match the SSC's quark power, the LHC would need more protons.

CERN had known this all along, and from the outset the aim had been to pump as many particles as possible into the LHC rings. It had been one of the reasons why a proton–antiproton collider option, with a single ring of magnets, had been ruled out; antiproton currents in the LHC could never reach the level required to ensure that enough quarks (and antiquarks) would reach 1 TeV, but there are limits as to how many protons can happily coexist in circulating beams. Put too many of them too close together and the two circulating beams start to interfere with each other. Instead of quietly circulating in the ring and only colliding at the required points, the beams become unstable.

The LHC design was carefully optimized to provide the high proton intensities, but these intensities were frightening—about ten thousand times higher than those used in the SPS proton–antiproton collider, a thousand times higher than the impressive levels attained in the ISR, and about ten times the level foreseen for the SSC. The full effect of such extreme proton levels had to fall on the LHC detectors, the giant assemblies intercepting the proton collisions and sifting carefully through the debris for examples of individual quark–quark or quark–gluon interactions deep inside the colliding protons. The detectors would have to be versatile enough to swallow a wave of data and at the same time be solid enough not to be destroyed by years of being buffeted by such waves. In an inferno of radiation, robust 'radiation-hard' equipment would have to pick needles out of haystacks. The research and development specialists had a job on their hands.

With the foundations taking shape, vertical political infrastructure needed to be developed. The LHC, with less time available than the SSC, had to get its experimental act together more rapidly. The SSC roadshow had first appeared in the mid-1980s, and the LHC too had to go out and bang the drum. In October 1990, some 500 physicists converged on Aachen, Germany, for an ECFA-sponsored LHC workshop. As well as the major sessions, physicists split up into groups for specialized workshops to study different aspects of LHC experiments: radiation resistance, simulation, software and data handling as well as the detector components themselves. From these workshops, the first nuclei of collaborations for LHC experiments emerged.

In December 1991, CERN Council moved that 'the LHC is the right machine for the advance of the subject and of the future of CERN', an encouraging amber light enabling the project to move ahead slowly. Until the scheme could be formally approved by CERN Council, the amount of investment and manpower that it could command was still limited. In March 1992, another LHC roadshow was staged at Evian, in France, on the south shore of Lake Geneva. Convenient because of its proximity to CERN, Evian was nevertheless far enough away from Geneva to encourage delegates to stay on site and to talk to each other in the evening rather than to go home. The attendance at Evian was both larger and came from farther afield than at the Aachen meeting. Aachen had been a strictly European event. For the first time, Evian reflected the world interest in the LHC. With more proposals for experiments on the table than the LHC could cope with, Evian also allowed physicists to regroup into larger collaborations. To push home the LHC message and to reach the scientists who could not make the Evian trip, subsequent 'mini-Evians' to explain the LHC idea and its research possibilities were organized in Russia, China and Japan.

From this round, four major detector projects emerged: Sam Ting's L3P scheme, using equipment and experience from the L3 experiment at LEP; Compact Muon Solenoid (CMS); ASCOT and EAGLE using superconducting magnets. Subsequently the latter two ideas joined forces as ATLAS, characterized by its choice of a toroidal geometry for the windings of its superconducting magnet, rather than using a conventional cylindrical solenoidal coil. In the toroidal approach, the coils are wound round a 'doughnut' geometry. Such magnet designs made their first impact in physics with the tokamak schemes to confine hot plasmas magnetically in experiments on controlled thermonuclear fusion. In the selection process for LHC experiments, ATLAS and CMS were the first major projects to emerge. These are vast undertakings, each as large as a five-storey building and weighing more than 10 000 tonnes, more than a fleet of Boeing 747 planes.

In October 1993, in the midst of LHC preparations, the news of the SSC cancellation hit CERN like a bombshell. People were confused and stunned. Instead of gloating, there was sympathy and offers of help. CERN management said, 'The European physics community hopes that collaboration can continue in the field which American physicists have been instrumental in creating and have led, with characteristic vigour, for several decades.' Europe silently recalled how Brookhaven, in 1952, had hosted a small group of CERN accelerator pioneers and magnanimously given them the gift of strong focusing—a new method for handling particle beams. Now it was Europe's turn to play godmother.

Although sidelined during the SSC effort, ICFA also stepped in. Policing the damage in the wake of the decision, ICFA alerted the international community on the need to focus on the LHC. At a meeting in Vancouver in January 1994,

ICFA, under its chairman, Fermilab Director John Peoples, drafted an important resolution which was to add additional weight behind the push for LHC approval by CERN Council. The resolution said, 'Following the cancellation of the SSC, the LHC at CERN now offers the realistic opportunity to study multi-TeV hadron collisions. ... The energy and luminosity of the LHC represent a great advance over the (Fermilab) Tevatron. There are compelling arguments that fundamental new physics will appear in the energy domain to be opened up by the LHC. ... ICFA believes that the time has come for the governments of all nations engaged in the science of high-energy physics to join in the construction of major high energy facilities, so that this unique human endeavour can continue to go forward.'

While international moves strived to repair the SSC damage, at a working level in the USA, the abrupt cancellation of the major project had caused near panic. As well as the employees of the Texas laboratory who found themselves without a job, hundreds of US researchers who had been working towards detectors for use at the SSC were cut off from a major research project. A tide of scientific talent immediately started to look for sanctuary. Ingenious solutions had been proposed for the SSC detector systems and it would be criminal to waste this effort. Prior to the SSC cancellation, the LHC with its lower collision energy naturally had been viewed in the USA as the outsider in a two-horse race. With only one runner, there was a rush to place bets before the race began. As well as the physicists looking to find the Higgs mechanism, there were the machine specialists, the successors of Ernest Lawrence and Robert Wilson. These accelerator physicists could help to build superconducting magnets, to orchestrate radio-frequency gymnastics, and to pack particles into beams. Here too was talent looking to be put to good use.

The US HEPAP, in its continual role of counselling the Department of Energy, quickly set up a post-SSC special 'future visions' subpanel chaired by the influential theorist Sid Drell of SLAC. The report said, 'We have inherited a great tradition of scientific enquiry. The field of particle physics has made dramatic progress in understanding the fundamental structure of matter. Recent discoveries and technological advances enable us to address compelling scientific issues ...' .

'As befitting a great nation with a rich and successful history of leadership in science and technology, the United States should continue to be among the leaders in the worldwide pursuit of the answers to fundamental questions of particle physics.'

'To sustain excellence in the US high-energy physics programme for two decades and beyond, three elements are essential,' the report continued. The first two elements were an effective ongoing national research programme based on upgrades of existing facilities, and vigorous studies for future accelerator and detector techniques. The third element was 'significant participation at the highest energy frontier, for which the best current opportunity beyond the (Fermilab) Tevatron is through international collaboration on the LHC at CERN'. The report went on to recommend increases in funding to accommodate these recommendations: 'The US Government (should) declare its intention to join other nations constructing the LHC at CERN and initiate negotiations toward that goal. Participation in the LHC should be endorsed with a timely

Figure 53: *Superconducting magnets for CERN's LHC under test.*

decision of support. This will enable the high energy community in the United States to take full advantage of this opportunity and to maintain momentum in the collaborations that have been forming in the hope of applying to the LHC the expertise and technology developed for the SSC and its detectors.' The writing was on the wall. Over the next few years, a wave of US interest engulfed the big ATLAS and CMS teams preparing big LHC detectors, to the extent that US physicists outnumbered those of any other single country.

The LHC was formally approved by CERN Council in December 1994. After such a long and arduous approval procedure, the LHC community breathed a collective sigh of relief. With the project formally blessed, potential interest in LHC from far afield became concrete support. CERN had formally courted these countries before, notably the USA, Japan, Russia, Canada, Israel, China and India but, with the project still falling short of European approval, intercontinental customers had been understandably reluctant to commit themselves. Now they were lining up to get in. The following CERN Council meeting, in June 1995, became almost a carnival after Japan gave five million yen for LHC construction. In acknowledgement of the generosity, Japan joined Israel, Russia, Turkey, the European Commission and UNESCO among the ranks of CERN 'Observers', who attend Council meetings but who cannot directly influence the decision-making process. In 1996, Russia also pledged substantial support. The Novosibirsk laboratory would supply some 500 magnets for about 5 km of beam lines to link the SPS with the LHC and through which the LHC will be fed with particles. Canada, India, Israel and the USA would also lend their support.

With such a level of intercontinental interest, the LHC, although managed in a European context, promises to become the first truly world particle physics machine.

FINANCIAL GIVE AND TAKE

A 'pork barrel' was proverbially a bountiful source, a bottomless well of plenty in times of adversity, but in US parlance the term has come to mean a project or operation which yields patronage benefits, such as employment or public spending, to a political constituency or its representative. The siting of the US SSC on a green field site in Texas is frequently cited as an example. From a scientific point of view, the optimum site for the US SSC would have been Fermilab, where the existing synchrotrons would have been able to launch the SSC protons into orbit and existing infrastructure could have been exploited. However, at the time, Texas was more adept at pulling strings than Illinois. The SSC would have provided a valuable intellectual focus for the culture-hungry Lone Star state, but the SSC's Texas label made it increasingly visible from Washington and hastened the project's downfall.

When CERN took advantage of Switzerland's offer of a home in Geneva, it was a compromise decision. Other, more prestigious, sites had been proposed, but the political connotations led to the low-key choice of Geneva, already home to several major international organizations, in a traditionally neutral country, but there was still a whiff of pork: such a vast new undertaking made a major impact on the local economy. While the scientists came from all over Western Europe, offices and workshops had to be staffed, and this new workforce predominantly came from the local region—Geneva and the neighbouring part of France. Unskilled workers suddenly found themselves with prestigious jobs. Geneva is a major city, a centre of communications, commerce and banking with its own newspapers and a university. Across the border in neighbouring France, the situation is very different. In this backwater, indigenous industry is traditionally sparse, and agriculture a major livelihood. Although in another country, Geneva had long been the economic focus of the estranged part of France east of the Jura mountains. Cut off from the rest of the country by the mountain range and by the river Rhône, the French *pays de Gex*, named after the principal town, has long-standing ties with Geneva. In the seventeenth century, the philosopher Voltaire, as a distinguished local resident, obtained from King Louis XVI of France important concessions on imports from what is now Swiss territory. Confirmed by several later decisions, the *pays de Gex* continues to enjoy special trade concessions with Switzerland, while the small canton of Geneva, with its limited population, cannot supply enough labour for an economy onto which has been grafted many large international organizations—the UN, the World Health Organization, the International Labour Organization, CERN, etc. Commuting backwards and forwards to work in Geneva became a regular feature of life in the *pays de Gex*. At CERN, initially sited at the French–Swiss frontier and subsequently straddling it, this tide of labour was especially visible. With CERN's finances pegged to the Swiss franc, the inexorable rise of the Swiss currency soon benefited the income of these commuters.

In the 1960s and 1970s, particularly with the decision to build the new Laboratory II, CERN's annual budget grew steadily. At the outset, the idea was that CERN contracts would be awarded on the basis of competitively tendering. However, when building the new CERN laboratory on Meyrin's green fields and in building subsequent large machines, Swiss construction firms won the lion's share of the work. The considerable concessionary powers of those in charge, the difficulties at the time of allowing foreign labour onto the site, and the legendary efficiency of Swiss operations all contributed. Over the period 1952–65, Swiss firms won some 35% of CERN business, although Switzerland only contributed some 3% of CERN's budget. Seen through Swiss eyes, CERN was an immense success, but not everybody finds pork to their taste. Within a decade, French firms had established a considerable presence and France, although initially disfavoured, went on to gain 20% of CERN's contracts, a figure comparable with the nation's contribution to the CERN budget. France got out what it put in. With specialist US suppliers taking another 13% of the business, this left only a third of the cake to be shared out among the remaining Member States, including Germany, Italy and the UK, who together provided more than half of CERN's funds.

As well as major contracts for equipment and supplies, the CERN labour force, which peaked at 3600 in 1975, spent its money locally. Especially in neighbouring France, housing estates and shopping centres sprang up to cater for this new wealth. CERN was not the only employer but was certainly a major one. Money from the coffers of London, Rome and Bonn was being spent in Geneva and Gex.

All this was only natural and initially in everyone's best interests, but other Member States began to cast envious eyes on this skewed distribution of CERN spending, and in 1992 a push began for 'fair return', that a determined effort be made to spread CERN contracts wider. As well as its considerable monetary value, much of CERN's business involved prestigious high technology which could seed new markets. A survey of economic spinoff for the period 1973–87 revealed that one Swiss franc spent by CERN in the high-technology sector went on to generate three times this value of new business. In the electronics sector, the build-up factor was more than four. The considerable value of CERN orders justifies research and development work which would otherwise be difficult to launch, and the prestigious contracts are a useful reference.

While the planet was already feeling the warmth of the greenhouse effect, the global economic outlook began to look chillier. At first the protected pastures of the Jura mountains and the tidy villages around Geneva did not feel this change in climate. However, the UK, exposed to the brunt of financial Atlantic currents, is a bell-wether for the rest of Europe. Already in the 1970s the UK had set a scientific precedent by closing major national accelerators and concentrating resources on CERN. In the early 1980s the Thatcher government took immediate action on the financial front, seeking to choke inflation and to limit public sector expenditure. The UK's reaction was at first incomprehensible to its European neighbours, for it took time for these chilly financial tides to sweep on. With the UK public purse tightly zipped, the size of the nation's contribution to CERN, then of the order of 50 million pounds annually, became highly visible. With only a limited research cake available, scientists in other sectors felt CERN's appetite was detrimental to their own interests. In March 1984, a committee chaired by Sir John Kendrew, who had won the Nobel prize for physiology and medicine in 1962, set out to investigate the UK's participation in high-energy physics in general and in CERN in particular. Its report concluded that, while the subject was worth pursuing scientifically, the proportion of funds it commanded was too high and advocated reducing the UK's contribution to CERN. It also threatened to 'reappraise' the UK's role in CERN—a euphemism for reducing the UK commitment or, as some thought, even pulling out altogether.

The report was widely criticized, and not only by particle physicists. *The Times* thundered, 'Withdrawal from CERN would be withdrawal from an unusually successful form of European joint action and of international collaboration. It would mark the end of Britain's long and leading contribution to the scientific study of the nature of matter. It would do more harm to the esteem and animal spirits of the scientific community in this country than any good the redistributed funds might do.' *The New Scientist* added, 'CERN's worth cannot be measured in Swiss francs, non-stick frying pans, or even in Nobel prizes. Its value lies in the fact that it has long surpassed its aim to bring Europe together. CERN has become a truly international laboratory.'

The Kendrew report fell on infertile ground but, if the report's objective had been merely to rock the boat, it succeeded. The furore that the report had created made it clear that CERN had to don sackcloth and ashes. A CERN review committee was set up under the distinguished French physicist Anatole Abragam and its report duly published in 1987. While the report spared no praise for CERN's scientific aims and achievements, it was clear that, at a financial and accounting level, new procedures were called for to bring CERN into line with current practice. Management had to be streamlined and some surplus fat shed.

In the meantime, CERN's budget came under pressure from an unexpected but nevertheless welcome quarter. The laboratory's scientific successes attracted more researchers; in the ten years from 1982 to 1992, this 'user' population climbed from 2500 to some 7000. While the laboratory congratulated itself on its new popularity, providing the research infrastructure to cater for so many extra researchers bit into CERN's budget, which had not been specially inflated to allow the big 27 km LEP electron–positron collider to be built. The 1300 million Swiss francs of construction money for LEP was eked out of running costs

During the Abragam committee deliberations, there had been discussions on ways of increasing CERN's income. One possibility was that the two 'Host States', France and Switzerland, who profited economically, should pay special additional contributions. However, the idea lay dormant until CERN Council began to discuss plans for the LHC proton collider, but now another force was acting. Instead of a financial current from the west, there were aftershocks following the political earthquake of perestroika in the east. The shaky edifice of communism collapsed, the Berlin Wall opened up and East Germany ceased to be a separate country. However much it boosted national pride, reunification in 1990 meant that Germany had to tighten its belt more than other nations. Absorbing what had been a separate country, with a population of some 16 million, made a big dent in Bonn's accounting. Germany quickly requested special treatment at CERN and, in 1993, CERN Council accorded the newly reunified country a five-year honeymoon period during which its contribution could be reduced by 10%.

These ill financial winds announced an icy winter which did nothing to entice the LHC into bud. In December 1991, despite being resolutely pushed by Director General Carlo Rubbia, CERN Council refused to budge, making instead a token acknowledgement that 'the LHC is the right machine for the future of CERN'. It was a useful rallying cry but was still a long way from project approval. To build the LHC would need some 2.6 billion Swiss francs, more than could be made available by drawing on CERN's normal operating budget. If the machine was to come into operation in the first few years of the twenty-first century, some 500 million Swiss francs of extra money would have to be found. Special contributions from major international physics players, particularly the USA, had long been seen as the source of this extra money, but these countries would not get their cheque books out before the LHC had been formally approved by its European partners, and the European partners would not approve the LHC before they had assurance that the extra money would be forthcoming. It was a classic chicken-and-egg situation.

In the 1980s, the CERN family had gradually enlarged. Spain, which had been a member from 1961 to 1968 and had then left, rejoined in 1983, giving 13 member states (Austria, Belgium, Denmark, France, Germany, Greece, Italy,

The Netherlands, Norway, Spain, Sweden, Switzerland and the United Kingdom). Subsequently, CERN welcomed Czechoslovakia (which later split into the Czech Republic and Slovakia), Finland, Hungary, Poland and Portugal. While this increased membership and natural expansion of scientific Europeanism was welcomed, it led to odd situations when votes had to be taken. Traditionally, voting rules required motions to be carried with at least a two-thirds majority, with unanimity being the rule for important decisions. With 19 Member States instead of 13, the four major contributors—France, Germany, Italy and the UK—could find themselves in a minority. To safeguard their interests and to prevent the financial fringe from dominating CERN affairs, an additional rule was adopted requiring that a two-thirds numerical majority should remain a majority when the votes were weighted by national contributions. In 1993, Germany, in dire straits in the wake of reunification, dusted off the old idea of host state contributions as a way to help the LHC. The UK, which had no reunification problem but was still keenly aware of the need to cut public spending wherever the opportunity presented itself, lined up behind Germany. This combined Germany–UK axis carried much weight.

In January 1994, Carlo Rubbia was succeeded as CERN's Director General by Oxford professor Christopher Llewellyn Smith. As well as being a distinguished theorist, Llewellyn Smith had played a key role in defending CERN's interests at the time of the Kendrew report and had instituted new methods for cushioning abrupt exchange rate fluctuations when doing CERN's budgetary calculations in Swiss francs. Under new management, CERN prepared itself for another attempt to push LHC approval through. Successive Council meetings had led CERN to pare manpower predictions and costs, with staff numbers foreseen as reaching a lower plateau at around 2300, more than a thousand down on the laboratory's 1975 complement of 3600. Fortunately the age profile of CERN's payroll meant that such a reduction could be achieved by natural or voluntary retirement.

Even after costs savings had been detailed, further cuts were demanded for and promised by 'somehow reducing support and services'. On 15 April 1994, a specially convened session of CERN Council confirmed its earlier declaration that the LHC was the right machine for CERN and said that its 'best intention was to move to a decision to approve the LHC during the first half of 1994'. With the first half of 1994 expiring only a few weeks away in June, the stage had been set for the traditional June Council meeting to approve the project formally.

Instead, the UK and Germany, citing increased domestic financial pressures, refused to budge and indeed threatened to veto the whole process unless new financial rules were introduced. At the same time, they increased pressure on France and Switzerland to make special concessions. These nations had already made generous gestures: France had given land and had led prestigious projects.

Switzerland had fostered CERN from the very beginning and in addition traditionally accorded CERN loans with low interest rates, but other nations wanted them to give more. When a motion to approve the LHC was put to the vote in June, 17 Member States voted in favour, but Germany and the UK abstained. According to the CERN rules, in principle this was enough to carry the motion, but in practice such a major project had to have unanimous support, and the motion was shelved for the subsequent meeting. Germany and the UK maintained their insistent pressure for tight budgeting and, as the December Council session drew near, threatened to torpedo the whole LHC approval process. With the pressure mounting, France and Switzerland dug deeper into their pockets and said that they would be willing to make special contributions for LHC construction, at the same time agreeing to pay 1% more for annual budget indexation above the level of the other Member States.

However, even building in all the cuts and the new money, and delaying the project by a few years, CERN's financial belt could not be buckled. A new option, carefully prepared behind the scenes, was proposed. The most expensive LHC item is its superconducting magnets. Instead of building the LHC with all its magnets operational from Day 1, a 'missing-magnet' scheme was proposed, in which the machine could initially be built with only two thirds of the required bending magnets. This would mean that the LHC would operate initially at a collision energy of 9.3 TeV (4.65 TeV proton beams), saving 300 million Swiss francs. This energy was too low to launch a serious attack on the Higgs mechanism but would nevertheless allow the LHC to make valuable introductory studies. In the following few years, the remaining magnets would be added. A similar missing-magnet scheme had been proposed when the SPS was in the design stage. When a question mark hung above funding for a new machine, any compromise was in danger of leading to a smaller lower-energy machine. In a missing-magnet scheme, instead of compromising the circumference of the ring, it is instead the magnetic bending power that is reduced. This 'cut-down' version could handle only lower-energy beams. However, if extra money were to become available later, this could be used to add more magnets to increase the bending power and attain a higher energy.

In the 1990s, major items of public expenditure became increasingly visible. In the USA, the SSC scapegoat was sacrificed on the financial altar, but Europe preferred across-the-board cuts to scapegoats. With public sector spending already pared to the bone in many countries, in February 1992 a new financial storm broke. European leaders signed the Maastricht treaty, pledging themselves to undergo financial purgatory to achieve economic unification. To pave the way for the introduction of a common currency, Maastricht stipulated that strict criteria had to be met. 'Member States shall avoid excessive government deficits,' says Article 104c. Specifically, budget deficit should not exceed 3% of gross domestic product, while national debt should be limited to

60% of GNP. Only Luxembourg could smile, and Luxembourg is not a CERN Member State.

For Germany, the cumulative effect meant something had to crack. In a financial drought, compelling arguments and fine words were useless. The ultimate criterion was what the government could pay. The German central bank, the Bundesbank, is normally concerned mainly with domestic matters, a heavy responsibility in the wake of German reunification, where Germany moved from a 3% surplus of savings over investments in 1990 to a deficit of 0.8% in 1995. Having built one of Europe's strongest currencies from the ashes of the war, Germany was now facing the prospect of having to abandon it in the face of the push towards an unpredictable common European currency—the 'euro'. In 1996 the Bundesbank uncharacteristically lifted its attention from domestic matters and exhorted other G7 countries to cut their budgets to reduce deficit spending and to avoid absorbing financial resources. German contributions to international ventures, including CERN, had to be reduced. This unilateral demand took many other countries, themselves committed to the same Maastricht criteria but seeking no such reduction in support for CERN, by surprise. A unilateral decision by one country to reduce its contribution was deemed unacceptable, but a compromise overall budget cut seemed inevitable.

On the sidelines, those who had grudgingly watched CERN devour cash gleefully rubbed their hands. Some scientists in other disciplines had envied CERN's ability to make ambitious dreams a reality. The proud gladiator was at last wounded and there was the scent of financial blood in the air but the wound, although serious, was not mortal. A compromise initial reduction of 7.5%, growing to 9.3%, meant that building the LHC would be even more of an uphill task, but there was good news, too. With LHC support from other nations, particularly Japan and the USA, and with Canada, India and Russia also making contributions, the intermediate-energy phase proposed in 1994 could be abandoned. The high road to LHC was open.

Ironically, CERN, one of the first major successes of European collaboration, was having to make sacrifices for the latest European collaborative venture—a common currency. CERN had shown how successfully individual nations can work together towards a collective goal when politics and national preoccupations are pushed aside. There is a European joke: 'Heaven is a place where the cooks are French, the mechanics are German, the police are British, the lovers are Italian, and it is all managed by the Swiss.' Hell is a place where the responsibilities are assigned differently, and the joke varies somewhat according to national tastes, but it has a ring of truth. CERN has demonstrated that, when objectives are goal oriented, differences can be overlooked and common interests reinforced, so that national characteristics dovetail smoothly together in a coherent effort. Managed correctly, Europe works.

FURTHER READING

Abraham Pais' *Inward Bound* (Oxford University Press) is a masterly account of the development of subatomic physics from the late nineteenth century to the mid-1980s, covering much of the same ground as this book. However, in places it is telescoped the wrong way in time, with historical episodes being covered in more detail than modern developments. This scholarly work also includes many equations and is aimed at professional physicists.

The following list is confined to books, and does not include learned journals or special publications with restricted circulation.

Exodus

Brown L M and Hoddeson L 1986 (ed) *The Birth of Particle Physics* (Cambridge: Cambridge University Press)

Crowther J G 1974 *The Cavendish Laboratory 1874–1974* (London: Macmillan)

Fermi L 1971 *Illustrious Immigrants, The intellectual migration from Europe 1930–41* (Chicago: University of Chicago Press)

McKay A 1984 *The Making of the Atomic Age* (Oxford: Oxford University Press)

Miller A 1984 *Early Quantum Electrodynamics, A Source Book* (Cambridge: Cambridge University Press)

There are many authoritative biographies, including the following.

Heilbron J 1986 *The Dilemmas of an Upright Man, Max Planck as Spokesman for German Science* (Berkeley, CA: University of California Press)

Kragh H 1990 *Dirac: A Scientific Biography* (Cambridge: Cambridge University Press)

Moore W 1989 *Schrödinger, Life and Thought* (Cambridge: Cambridge University Press)

Pharr Davis N 1968 *Lawrence and Oppenheimer* (New York: Simon and Schuster)

Wilson D 1983 *Rutherford* (London: Hodder and Stoughton)

The storm breaks

Brown L M and Hoddeson L (ed) 1983 *The Birth of Particle Physics* (Cambridge: Cambridge University Press)

196

Feynman R 1985 *Surely You're Joking, Mr Feynman* (New York: Norton)

Gleick J 1992 *Genius* (New York: Little, Brown)

Gowing M 1964 *Britain and Atomic Energy 1939–1945* (London: Macmillan)

Hanbury Brown R 1991 *Boffin* (Bristol: Institute of Physics)

Jones R V 1978 *Most Secret War* (London: Hamish Hamilton)

Lovell B 1991 *Echoes of War* (Bristol: Institute of Physics)

McKay A 1984 *The Making of the Atomic Age* (Oxford: Oxford University Press)

Ng Y J (ed) 1996 *Julian Schwinger, The Physicist, The Teacher, and the Man* (Singapore: World Scientific)

Rhodes R 1986 *The Making of the Atomic Bomb* (New York: Simon and Schuster)

The grapes of wrath

Brown L, Dresden M and Hoddeson L (ed) 1989 *From Pions to Quarks: Particle Physics in the 1950s* (Cambridge: Cambridge University Press)

Foster B and Fowler P (ed) 1988 *40 Years of Particle Physics* (Bristol:Adam Hilger)

Gleick J 1992 *Genius* (New York: Little Brown)

Ng Y J (ed) 1996 *Julian Schwinger, The Physicist, The Teacher, and the Man* (Singapore: World Scientific)

Rabi I 1960 *My Life and Times as a Physicist* (Claremont, CA: Claremont College)

All we want is the world's biggest machine

Crowley-Milling M 1993 *John Bertram Adams, Engineer Extraordinary* (New York: Gordon and Breach)

Hermann A, Krige J, Mersits U and Pestre D 1987 *History of CERN* (vol 1) (Amsterdam: North Holland)

Waloschek P 1994 *The Infancy of Particle Accelerators, Life and Work of Rolf Wideröe* (Braunschweig: Vieweg)

The quark in the box

Close F, Marten M and Sutton C 1986 *Particle Explosion* (Oxford: Oxford University Press)

Gell-Mann M 1994 *The Quark and the Jaguar* (London: Little Brown)

Matthews P 1971 *The Nuclear Apple* (London: Chatto and Windus)

Riordan M 1987 *The Hunting of the Quark* (New York: Simon and Schuster)

Sutton C 1992 *Spaceship Neutrino* (Cambridge: Cambridge University Press)

Glued from the spot

Close F 1989 The quark structure of matter *The New Physics* (ed) P Davies (Cambridge: Cambridge University Press)

Fritzsch H *Flucht aus Leipzig* (München: Piper)

Fritzsch H 1983 *Quarks* (London: Allen Lane)

Ideas in collision

Breizman B and Van Dam J G 1994 *I Budker, Reflections and Reminiscences* (New York: American Institute of Physics)

Crowley-Milling M 1993 *John Bertram Adams, Engineer Extraordinary* (New York: Gordon and Breach)

Hermann A, Krige J, Mersits U, and Pestre D 1990 *History of CERN* Vol II (Amsterdam: North-Holland)

Weisskopf V 1991 *The Joy of Insight* (New York: Basic Books)

Vacuum-packed physics

Galison P 1987 *How Experiments End* (Chicago, IL: University of Chicago Press)

Glashow S 1988 *Interactions* (New York: Warner Books)

Lai C H (ed) 1984 *Ideals and Realities, Selected Essays of Abdus Salam* (Singapore: World Scientific)

Lundqvist S (ed) 1992 *Nobel Lectures 1971–80* (Singapore: World Scientific)

Newman H and Ypsilantis T (ed) 1996 *History of Original Ideas and Basic Discoveries in Particle Physics* (New York: Plenum)

Riordan M 1987 *The Hunting of the Quark* (New York: Simon and Schuster)

Rousset A 1995 *Gargamelle et les Courants Neutres* (Paris: CNRS)

Weinberg S 1995 *The Quantum Theory of Fields* (vols 1 and 2) (Cambridge: Cambridge University Press)

Physics from AA to Z

Fraser G, Lillestøl E and Sellevåg I 1994 *The Search for Infinity* (London: Mitchell Beazley)

Krige J 1996 *History of CERN* vol III (Amsterdam: North-Holland)

Taubes G 1986 *Nobel Dreams* (New York: Random House)

Watkins P 1986 *The Story of the W and Z* (Cambridge: Cambridge University Press)

INDEX